ANIMAL BIODIVERSITY : PATTERNS AND PROCESSES

Prof. T. N. Ananthakrishnan
Emeritus Scientist
and
Prof. K. G. Sivaramakrishnan
(Former Professor and Head,
Department of Zoology, Madura College, Madurai)

SCIENTIFIC PUBLISHERS

Published by
SCIENTIFIC PUBLISHERS (INDIA)
5 A, New Pali Road, P.O. Box 91
Jodhpur 342 001 (INDIA)
E-mail: info@scientificpub.com
Website: http://www.scientificpub.com

ISBN: 979-81-72334-30-7
eISBN : 978-93-87307-47-6

Printed in India

Ecosystems of India

PREFACE

An increased awareness of the impact of science and technology on society over the last two or three decades have led to an increased realization of deterioration of biological systems, which has assumed global dimensions. Technology has resulted in ecological changes at a very fast pace, and with our limited understanding of the functioning of ecosystems, the intensity of the implications pertaining to the loss of biodiversity i.e. the loss of species, genes and ecosystems have not been registered.

Immense economic benefits accrue to man because of the prevailing biodiversity. Therefore continued improvement of biodiversity, and hence the benefits, will depend on new and enhanced resources from nature. At any given time, changes in biodiversity, i.e. the increase or reduction or maintenance of the diversity of genes, species or ecosystems will depend largely on human activities. Access to these resources will therefore depend on scientific knowledge of these resources through studies of biodiversity to enable prediction of the most promising species, and choosing sites for prospective biological resources, which in turn will provide relevant information from the countless number of species.

The identification, recognition and emphasis related to this multifaceted discipline have assumed an increased relevance today, especially when such issues as environment, energy, global changes and sustainable development have become a part of basic education elsewhere in the world, aiming at an increased integration between basic and applied sciences. Furthermore, because of socio-economic changes, biological diversity has today come to occupy the central stage as it holds the 'key to the maintenance of the world'. It has emerged as a unifying discipline bringing together the ecologist, environmentalist, educationist and economist, resulting in an interdisciplinary, multifunctional, problem-oriented education. Thus the essence of this education emphasizes the relevance and quality to cope with issues like ecosystem dynamics, environment and climate changes, energy sources, biotechnology, global changes, and sustainable development at the local, national and global levels.

Attitudes towards the diversity of species in diverse ecosystems are also changing as people face the impact of the direct cost of the loss of genes, species and ecosystems. Conservation is therefore increasingly recognized as essential to economic development. Recent attempts in this regard have resulted in saving biodiversity, studying it and using it sustainably and equitably. The need to understand and validate traditional ecological knowledge for managing biodiversity by the local people has also come to be appreciated. Sustainably managing our biosphere in the face of global change is the need of the hour.

This book therefore attempts to provide an overall emphasis of diverse aspects of animal biodiversity involving all aspects referred to above. Brief extracts from the senior author's earlier publications on *General Animal Ecology* (Macmillan, 1985) and *Bioresources Ecology* (Oxford and IBH, 1985) on soil and aquatic ecosystems have enabled a better integration of the diverse aspects discussed.

In the preparation of this volume, we wish to express our thanks to Dr. K. Venkataraman of the Zoological Survey of India, Chennai, for his expertise and assistance, in particular on marine ecosystems and for providing some of the illustrations. To Dr. V.V. Ramamurthi of the Indian Agricultural Research Institute, New Delhi, Dr. B. V. David of Sun Agro Biotech Research Centre, Chennai, Dr. R.V. Varma of the Kerala Forest Research Institute, Peechi, Dr. S. Uthamasamy of the Tamil Nadu Agricultural University, Coimbatore, Dr. Sabu Thomas of St. Joseph's College, Calicut and Dr.V. Oommen of the Department of Zoology, University of Kerala, Trivandrum, we are indebted for their ready response to our request for some photographs. Thanks are also due to Shri E. Ragupathy for his painstaking efforts not only in typing the manuscript, but also scrutinizing it to be as error free as possible and to Mr. Jahir Hussain for his assistance in typing the sections dealing with literature, glossary and Index. We wish also to acknowledge the interest evinced by the Publishers, in the production of this volume.

Chennai
1.11.2005

T.N. Ananthakrishnan
K.G. Sivaramakrishnan

List of Illustrations

CONTENTS

CHAPTER - 1

BIODIVERSITY : DIMENSIONS AND IMPLICATIONS

An increased awareness of the impact of science and technology on society over the last few decades has led to an increased realization of deterioration of biological systems, which has assumed global dimensions. Increased human interference has threatened the stability of many ecosystems – the terrestrial and aquatic – reducing areas of habitats, resulting in ecological changes at a very fast pace with consequent reduction of species. With our limited understanding of the functioning of ecosystems, the implications pertaining to the loss of species, genes, communities and ecosystems are indeed great. Reduced populations make species vulnerable to disease, pollution, habitat alteration and inbreeding, with confounding effects of the impact of climate change and global warming. This disappearance of species of plants and animals as a result of human activities throughout the world has been a cause of great concern. It is in this context that a group of committed environmentalists and ecologists coined the term biodiversity. Biological diversity was first proposed by Norse and McManus (1980) for two combined concepts involving genetic diversity and ecological diversity. While the former relates to the amount of genetic variability within species, the latter refers to the number of species in a community of organisms. The term 'biodiversity' was accepted at the first planning meeting of the National Forum on Biodiversity in Washington D.C. in 1986. Biodiversity includes all life forms, ecosystems and ecological process at the genetic, taxonomic and ecosystem levels (McNeely *et al.*, 1990). The National Academy Press published the proceedings of the forum in 1988 under the title 'Biodiversity', an alternate word for biological diversity and cited in subsequent publications as biodiversity (Wilson and Peter, 1988). During the Earth summit held in Rio de Janiero in the summer of 1992, biodiversity was a key topic of global, scientific, economic and political concern. In all its manifestation, it is an essential component of the quality of human existence, so nicely epitomized in the ancient aphorism, 'variety is the spice of life'.

Definition and Scope

The classical definition of biodiversity included in the convention on Biological Diversity (UNEP, 1992), defines biodiversity as "the variability among living *inter alia* terrestrial, marine and other aquatic systems and the ecological complexes of which they are part; this includes diversity within species, between species and of ecosystems". According to Global Biodiversity Strategy, biodiversity is "the totality of genes, species and ecosystems in a region". While Huston (1994) views biological diversity, essentially as "the coexistence of species on changing landscapes", Wilson (1997) presents biodiversity as "all heredity based variation at all levels of organization, from the genes within a single local population or species, to the species comprising all or part of a local community and finally to the communities themselves that compose the various ecosystems of the world". Mallet (1998) argued that diversity of life is fundamentally genetic and one can use a variety of genetic mutations to investigate diversity both within and between species.

Plants and animals continue to coexist with us in this planet, where we have most of our interactions relating to biological diversity, through appreciation of the common, strange and beautiful forms of life. The potential utilitarian, economic and aesthetic values or through suffering caused by the effects of pests, parasites, predators and disease transmitting vectors, have engaged us in biodiversity studies right from Linnaean times under the garb of taxonomic explorations. However what has given it such extraordinary, widespread attention is the present '*biodiversity crisis*' caused by the alarming rate of the disappearance of species, the so called '*biodiversity erosion*' from the face of the earth.

The determination of the components of biodiversity in a region, their distribution and interrelationships, their possible threats, techniques of their measurement and short-term and long-term monitoring methods and formulation of pertinent conservation strategies relevant to respective habitats, biomes and biogeographic zones, have become very important. Biodiversity is considered as "the variety of life and its processes, including the variety of living organisms, the genetic differences among them, the communities and ecosystems in which they occur, and the ecological and evolutionary processes that keep them functioning, yet ever changing and adapting".

The four levels of biological organization usually identified are genetic, population, species, community, ecosystem and landscape or regional. Each of these levels can be further divided into compositional, structural and functional components of a nested hierarchy (Noss, 1990). Conserving biodiversity involves not only saving species from total extinction, but taking steps to counteract biotic impoverishment which can take many forms and occur at various levels of biological organization.

Genetic Diversity

Species and their population contain a precious genetic library, the most important endowment of biodiversity, maintained by natural ecosystems. Genetic variation within the species influences the way in which a species interacts with its environment and enables a population to respond to natural selection. Genetic diversity also provides the potential for subsequent evolutionary change. Biodiversity among organisms ultimately arises at the molecular level involving the sequences of nucleic acids (Singh and Khurana, 2002). Genes, which are sequences of DNA molecule (deoxyribonucleic acid) are the functional units of heredity. Genetic diversity refers to the diversity (or variability of genes) within species. The terms also cover distinct populations of a single species. Random changes like mutations, recombination and genetic drift, besides changes like natural and artificial selection promote genetic diversity. This is fundamental to the variety of life and is the raw material for the evolution of new species, which is a less obvious level of diversity when compared to species diversity. Conservation goals at the genetic level include maintaining genetic variation within and among populations of species and assuring continuation of genetic differentiation and gene flow at reasonably normal rates. Domestication of certain species of wild animals by artificial selection and creation of new varieties of 'transgenic' plants and animals by means genetic engineering are aspects of promoting genetic diversity. However greater possible caution should be exercised in this area, so as to avoid distortions and freaks, which may have unforeseen consequences.

Species Diversity

Species diversity, in its broadest sense refers to the numbers of species occurring in a particular location. Species diversity is often measured with the goal of examining and comparing patterns of species distribution at local and regional levels. These measures are used chiefly for examining particular groups of species rather than the full range of species and interactions found in nature.

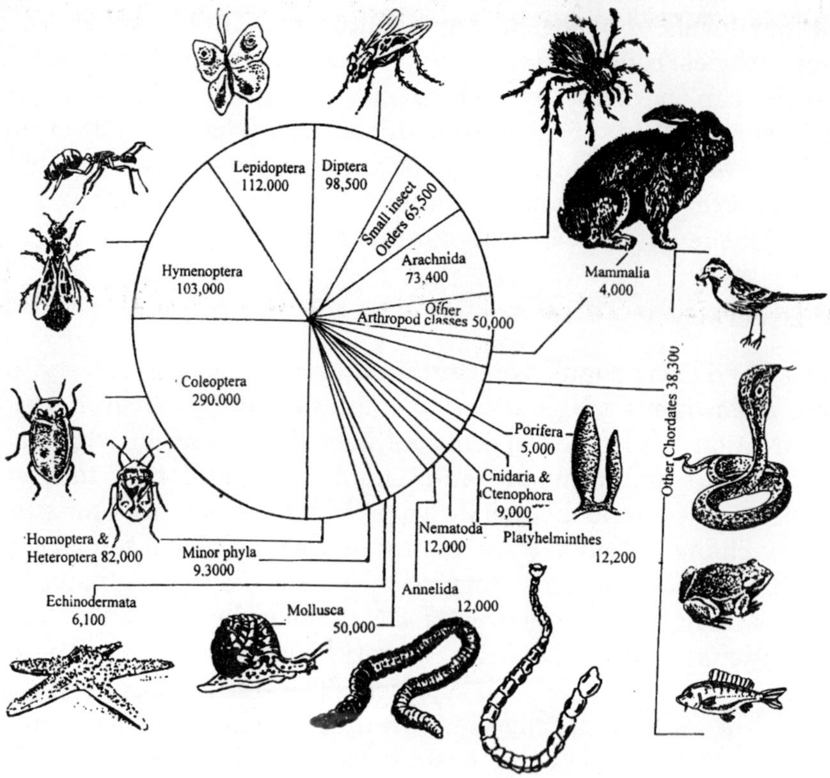

Figure 1.1. Categories of Biodiversity

Species diversity has mainly three components, namely *species packing*, *species turnover* and *geographical turnover*. The diversity of living organisms is not evenly distributed on earth. The categories of species that are classified, named, declared and included under biodiversity are represented in Fig. 1.1. On land, the humid tropical habitats are far more diverse. In the deep sea, the cold, unchanging depths of oceans are especially rich. The much greater variety of tropical life has been attributed to a variety of reasons such as greater energy, stability with increased solar energy, productivity, year round occurrence of conditions, climate favourable to life and greater biomass production. All these factors have led to higher levels of *species packing* in the tropical latitudes. *Species turnover* or the replacement of one set of species by another is the second component of species diversity. Species turnover levels are particularly high at the marine intertidal zones and mountain tracts where environmental regimes change rapidly. The two mountain regions *viz.* the Western Ghats and Eastern Himalaya figure among the world's 25 "*hot spots*" *of biological diversity* which are areas of high species richness or of high endemic species, thus claiming high priority for conservation (Fig. 1.2). The Eastern Himalayas with some 3,500 endemic species of higher plants, 20 endemic species of reptiles and 25 endemic species of amphibians, and the

Western Ghats with 1,600 endemic species of flowering plants, seven endemic species of mammals, 91 endemic species of reptiles and 84 endemic species of amphibians (Gadgil, 1997), have qualified for the status of hot spots. India has been identified as one of the top megadiversity countries and ranks eleventh in terms of the total number of species of endemic reptiles, birds and mammals. *Geographical turnover*, the third component of species diversity is the proportion of species unique to a region. Species restricted to a given region are said to be endemic to that region. The lion-tailed macaque in southern Western Ghats is a classical example. Islands by virtue of their long isolation are especially rich in endemic species. For instance, Australia has maximum numbers of endemic species as of mammals (210) and reptiles (605). 18 richest multitaxon centres of endemism were recognized by Robert *et al.* (2002) in the marine realm, identifying 10 marine biodiversity hotspots on the basis of endemism and threat. These cover 15.8% of the world's coral reefs.

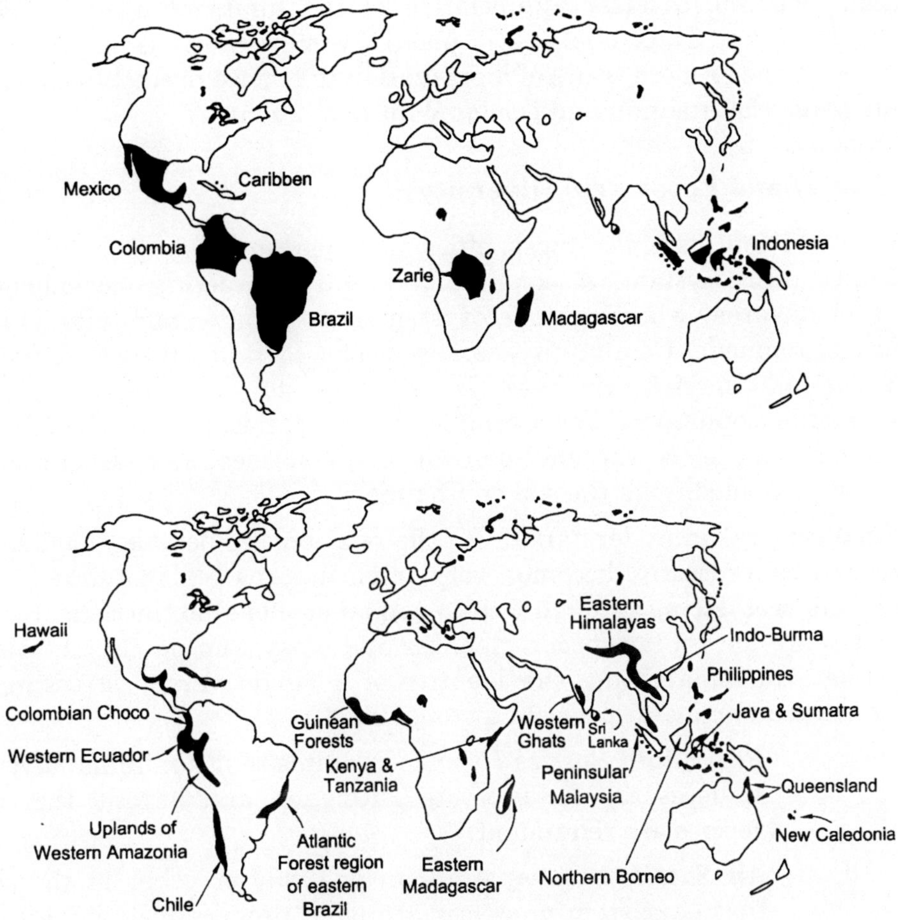

Figure 1.2. Biodiversity rich countries (Top) : Hotspots of Biodiversity (bottom)

Phenotypic Plasticity

Intraspecific diversity is an essential component of biodiversity and species living in heterogeneous environments show polymorphism or phenotypic variations, which are reflected in populations as *phenotypic plasticity*. The degree of phenotypic plasticity required to provide a range of variation can only occur if there are sufficient genetic variants. Such a plasticity ensures survival and reproduction under any set of environmentally induced, adaptive, behavioural, physiological or ecological variations among individuals of a population resulting in phenotypic plasticity. Gene flow between habitats significantly elevates variations and in heterogeneous environments, there are adaptive advantages that allow environmentally induced expression of phenotypes. In other words, phenotypic variation can be viewed in terms of the realized manifestation of multiple phenotypes within a population. An assessment of the magnitude of such variations becomes essential for the proper recognition of the patterns of variation, enabling a letter appreciation of the limits of diversity of the concerned species. Thus phenotypic plasticity may range from continuous variations in one or more traits of a population to a more predictable fixed set of traits (Ananthakrishnan and Doglas Whitman, 2005).

Community and Ecosystem Diversity

Conservation becomes most efficient when focused directly on the community or ecosystem. A community is an interacting assemblage of species in an area, whereas an ecosystem is a biotic community plus its abiotic environment. Community ecology seeks to find an explanation for differences in species richness in relation to differences in local environmental conditions. For example, some species modify the physical environment and hence the environment of the species, so characteristic of earthworm taxa modifying the soil environment.

Furthermore, an understanding of the relationship between biodiversity and ecosystem processes becomes very important as for instance, loss of species from ecosystems through local, regional or global extinctions. Several hypotheses are known for species richness and ecosystem processes, some of which relate to highlighting the identity of a single versus stressing the importance of an array of species (Lawton, 2000).

a. *Redundant Species Hypothesis* states that a minimal diversity is necessary for ecosystem function and beyond this most species are redundant.

b. *All Species are important* in one way or other in the sense that ecosystem processes are more progressively impaired as species are lost from the ecosystem.

c. Several species may be important in ecosystem processes, but *identity of particular species and its traits matter more* than species richness with some species contributing more to the ecological integrity of the ecosystem, than others.

An understanding of the regulation of species diversity in an ecosystem is made possible when the species forming a community is classified based on "functional" attributes; that is, attributes related to how organisms interact with each other and with their environment, rather than on phylogenetic attributes and genetic relatedness. According to this approach, an ecosystem consists of an assemblage of structural and inferential species like oysters, mussels, barnacles, tubeworms, etc. Examples of s*tructural species* are reef-forming corals which create or provide the physical structure of the environment, producing variability in physical (e.g. microclimatic) conditions, providing resources and in general, creating the habitat used by many others generally smaller, '*interstitial*' organisms. Interstitial organisms would include insects, other arthropods, birds, mammals and other vertebrates, microbes and fungi as well as plants such as epiphytes and understorey herbs. Structural organisms have a major influence on the diversity of interstitial organisms. In most cases the interstitial organisms would not be present in a particular area without the structural organisms. However, the direct influence of interstitial organisms on structural organisms and their diversity is usually minor, although the indirect effects of interstitial organisms through evolutionary and biogeochemical processes can be very important.

The total diversity of any community can be broken into two hierarchical components: one is the number of different functional types of organisms forming '*species guilds*', which are taxonomically different, but ecologically identical species ecosystems flourishing at great heights in the forest canopy, often described as the 'undiscovered continent' or `power houses of the forests'. They harbour species guilds in countless numbers, best known for the ecological services they offer, especially in the form of pollinators and bio-control agents. Incorporating species into ecosystem function results in functional groups.

The second component is the number of *functionally analogous species* within each functional type. Different functional types of organisms use different resources. They interact with their environment in such a way that competitive interactions are minimized. Species of the same functional type use the same resources in such a way that they could potentially compete intensely with one another. Functionally analogous species may result from convergent evolution in which distantly related organisms become adapted to similar environmental conditions. A good example is the adaptive radiation of the marsupials of Australia, comparable to the placental mammals of the rest of the world. From an ecological perspective, interactions such as competition or predation/commensalism among the very similar species within a functional type are likely to operate with difficulty, than the same classes of interactions among species of different functional types. Patterns of species richness relate to such diverse factors as an ancestral species splitting into two or more daughter species; changes in the geographic ranges by range movements and changes occurring in the genotype and phenotype of species. Competition and predation also tend to reduce diversity through elimination of taxa. Ecological diversification is also important since taxa originate and diversify in same habitat types (Barraclough *et al.*, 1998).

Thus, an ecosystem is a dynamic unit in which the competition, structure and roles of the different elements within it are constantly evolving with time. The role of biodiversity in the functioning of different ecosystems has not been explored thoroughly. Practising conservation at the community or ecosystem level demands attention to ecological processes, which affect biodiversity at several levels. In terrestrial communities some of the most important processes are fire and other natural disturbances, hydrological cycles, nutrients cycling, plant-herbivore interactions, predation and soil building process. To keep native biodiversity at reasonable levels, these processes must be maintained at normal limits of variation. Human induced disturbances like logging and livestock grazing under the garb of forest management should not interfere with basic ecological processes, disrupting nutrient retention. A major future challenge is to determine how biodiversity dynamics, ecosystem processes and abiotic factors interact, since a large number of species are probably needed to reduce temporal variability in ecosystem processes in changing environments (Fig. 1.3).

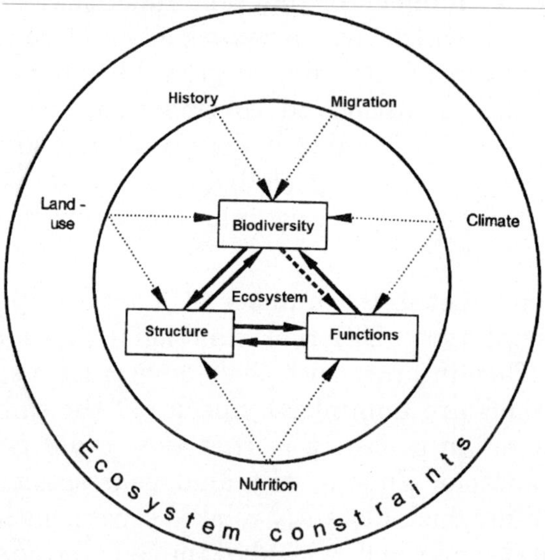

Figure 1.3. Interaction of Biodiversity, ecosystem process and abiotic factors.

Bioinvasions

The colonization by exotic species in an alien environment has introduced considerable problems and the well known invasion of water hyacinth in our aquatic system and of lantana invasion on land are notable examples. While human mediated large scale movement is the cause of this biological invasion, advancing technological breakthrough in the area of transport and communications has converted the world into a 'global village', besides land transformation which provides ideal ecological conditions for the invasive species to colonize and expand. The end result is land degradation and all that is associated with it, including rapid depletion of biodiversity, calling for consideration of an integrated landscape management approach towards managing biological invasion (Ramakrishnan, 2004).

These global changes besides causing *biodiversity depletion*, including species or subspecific level losses, because of habitat fragmentation, impact of invasive species and climate change impact, also result in alterations in ecosystem structure and function. Biodiversity is being lost alongside with its many economic benefits as food, fibre, medicinal plants etc. even before the biota is identified and catalogued. The buffering capacity of the ecosystem and its adaptability to cope up with uncertainties in the context of global change are seriously affected because of the large scale depletion of biodiversity.

Ecological indicators

Land use and land cover changes, biomass quality and quantity, water quality and quantity and soil fertility and energy management are essential aspects of sustainable development. Sustainable management of soil fertility implies the effectiveness with which it is possible to maintain soil fertility by strengthening internal biological processes within an ecosystem (Ramakrishnan, 2004). The quantity of biomass generated (biodiversity) on the land and the quantity of biomass generated through rehabilitation measures of a given ecosystem, are important indicators.

Traditional Ecological Knowledge (TEK)

In recent years, traditional wisdom based on the theme that there should be a close partnership between man and nature has come to be increasingly recognized. Is the context of global change, traditional ecological knowledge encompassing all aspects related to ecology and natural resource management has assumed importance. The threefold benefits accorded for their knowledge necessarily relates to economic, ecological and social and ethical aspects. Particular reference may be made to sacred groves and sacred landscapes, which are ethical; traditional crop varieties and lesser known plant and animals of food value and medicinal plants which account for economic aspects; and utilization of biodiversity for coping up with global changes as well as for soil and water management. The accompanying chart indicates the interaction between ecological and social sciences, in terms of biodiversity, land use and land cover dynamics (Ramakrishnan, 2004).

Landscape and Regional diversity

"Landscape is a heterogeneous area composed of a cluster of interacting ecosystems that is repeated in similar form throughout" (Forman and Gordon, 1986). In other words, it is a "mosaic of heterogenous land forms, vegetation types and land users" (Urban *et al.*, 1987). Landscapes do have a pattern which consists of repeated habitat components that occur in various shapes, sizes and spatial interrelationships. In many landscapes, this pattern consists of patches and corridors in a *matrix,* which is the most common or interconnected habitat in the landscape.

The concept of landscape diversity also takes into account the relationships between landscapes viz. their territorial organization, dynamics and inter-relationships as seen by individuals and societies through different local, regional and national cultures. The quality and diversity of a landscape are the products of human activities over thousands of years. They are continually evolving owing to the constant changes in the way that different societies use land.

Ecoregions are large landscapes that can be distinguished from other regions on the basis of climate, physiography, soils etc. *Pattern diversity* involves species composition patterns and other variables, landscape or regional diversity. It is the pattern of habitats and species assemblages across an extensive land area. This can be considered as a higher level expression of diversity. Loss of biodiversity in ecosystems and landscapes have important implications, including dimension and resistance to disturbance, system simplification and loss of ecological integrity. Certification to preventing losses of biodiversity is an understanding of patterns of *taxa richness*, a commonly used measure of biodiversity on a variety of *spatial scales*. However, our knowledge of broad-scale patterns of many ecosystems is especially meagre.

It is in this context that Huston (1979) has come out with a dynamic equilibrium model of species diversity, incorporating many of the major concepts in ecology into a framework for understanding patterns of diversity on landscapes. A primary conservation goal at the landscape regional level is to maintain complete, unfragmented, environmental gradients. This is because fragmentation of habitats across environmental gradients leaves remnants no longer connected to a larger wilderness and hence species are lost over time.

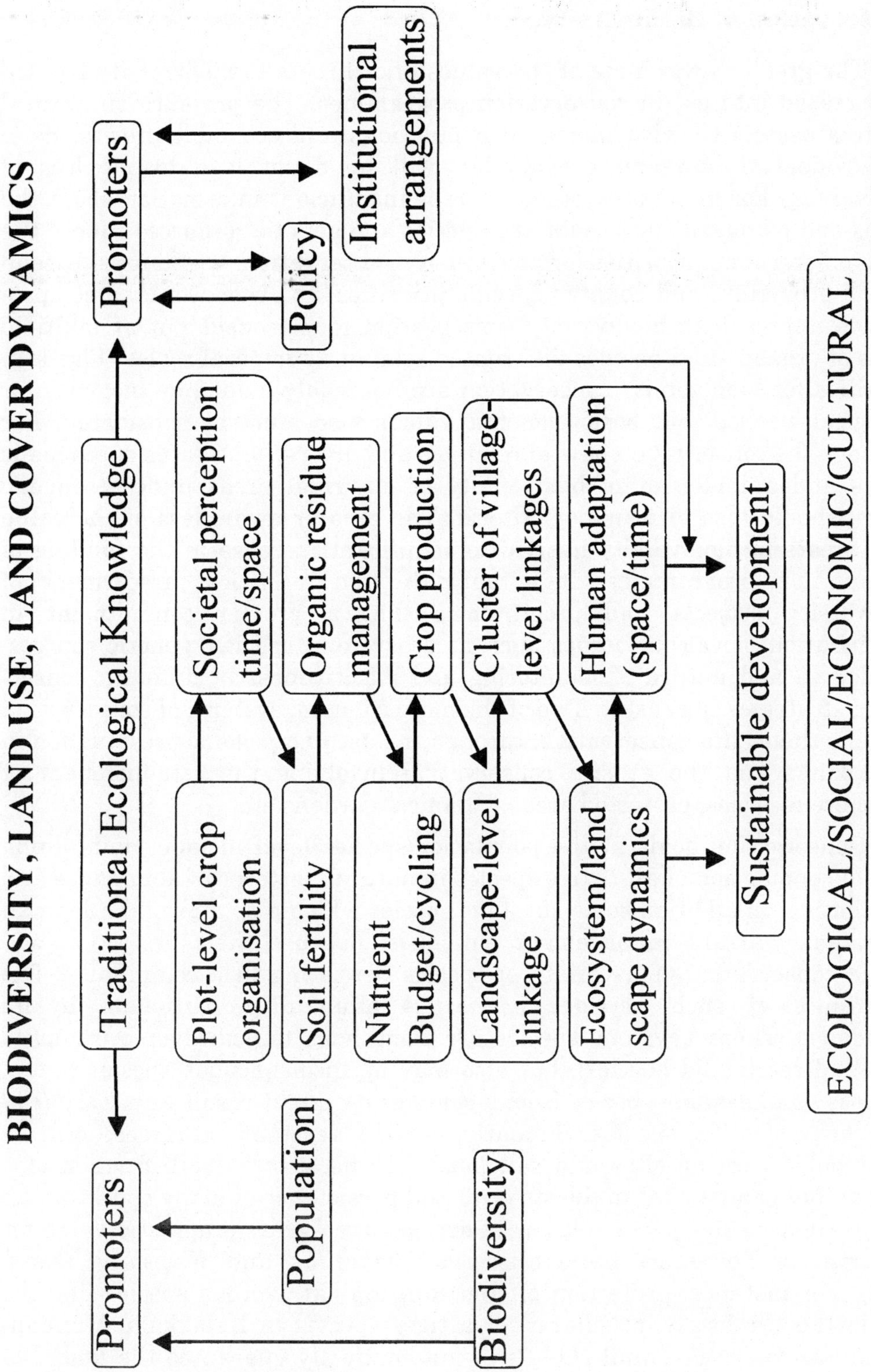

(After Ramakrishnan, 2004)

Figure 1.4

Conservation of Biodiversity

The greater awareness of the values and threats to biodiversity has led to increased interest in conservation programmes. The pressure on natural habitats associated with increased populations and economic growth, as is well evident in developing countries, will no doubt lead to the loss of biodiversity. For many ecosystems there is insufficient information about the status and management of habitats, species and genetic resources. Hence the need to determine priorities for biodiversity conservation, which must extend beyond ecosystem and country specific priorities. In other words, a comprehensive approach to biodiversity conservation must consider it at multiple levels of organization besides different spatial and temporal scales (Fig. 1.4). Priorities for biodiversity conservation are not solely guided by biological or ecological criteria, but socio-economic values also must be considered. For example, the conservation of wild relatives and diverse land races of economic plants and animals, or of biodiversity of marginal areas under economic stress, should also form major criteria. The greater awareness of the value and threats to biodiversity has led to significant increase in the funding of conservation programmes. It is only through a critical assessment of biodiversity projects and programmes that a greater commitment to prioritization of various issues such as ecosystem, species, genetic sources, the degree and nature of monitoring and effectiveness of different conservation strategies can result. Two of the more common systems of conservation relate to the *in situ* conservation, through the use of cryogenic preservation in gene banks and the *ex situ* conservation involving the establishment of biosphere reserves, protected areas, botanical gardens etc.

Conservation goals at the population/species level include maintaining variable population of all native species in natural patterns of abundance and distribution. In addition some keystone species which play a pivotal role in an ecosystem, warrant management emphasis, because their protection will assure conservation of a variety of species persisting in a community. The importance of such *keystone species* is often disproportionate to its abundance. These keystone species are sometimes top carnivores (reduviid bugs and coccinellid beetles), but also may be inconspicuous species (ants). The loss of a keystone species from a community might result in a *cascade of extinctions* of other species. Similarly, certain keystone resources, such as water holes, nesting sites and salt licks may be a very small fraction of a habitat, but can be vital to the survival and persistence of many species in an area (Primack, 2002). Some kinds of species have great pragmatic value for conservation. These are characterized as "*Umbrellas*" and "*Flagships*" (Noss, 1990). For instance, protection offered to a top carnivorous species like the lion in the Gir forest in Gujarat, or a tiger species in Kalakkadu-Mundanthurai Sanctuary in Tamil Nadu will automatically guarantee the umbrella of protection to a variety of less-demanding species inhabiting those vast stretches of natural forests. Umbrella species are often charismatic (Asian elephant, for example) and hence they also function as *flagships* or symbols for major conservation efforts (Noss and Cooperrider, 1994).

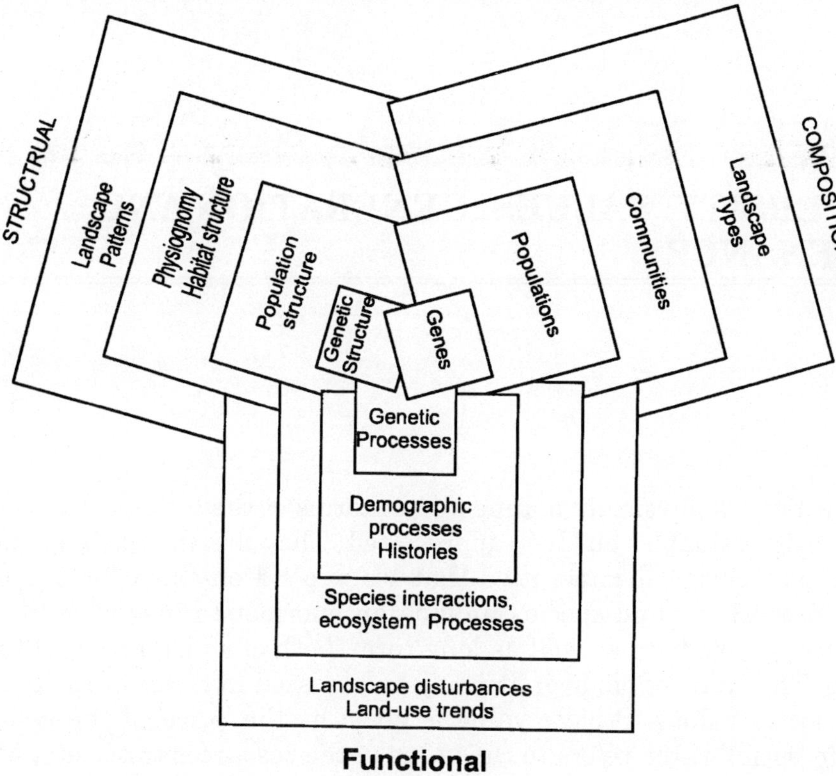

Figure 1.5. Biodiversity at multiple levels of organization (After Noss, 1990)

A more detailed information in the conservation of biodiversity and implications involved, is presented in Chapter 17.

CHAPTER - 2

BIODIVERSITY: VALUES, GENERATION AND MAINTENANCE

Biodiversity is physically manifested in terms of the biological resources which are most valuable, but less appreciated. They are the basis of life on earth. It is an ecological imperative that we depend on them for our basic existence. Ecosystems and species provide an enormous range of goods and other services–immediate as well as long term, both of which are vital to our good living. The value of biodiversity may be assessed in terms of its 'private' values or 'social' values. Private value is given by the 'price' of the resource itself, while 'social' value refers to the value of the resources to society, which varies in countries with different cultures. Social economic status, social performances, degree of technological advances and the purchasing power of a nation's currency are largely responsible for the differential perceptions of the value of bioresources between countries. Private values of biodiversity components are to be analyzed in order to find the driving forces behind diversity loss. An understanding of the social value of biodiversity is also essential to ascertain how much biodiversity should be conserved (Perrings, 1995).

Biodiversity is useful to humanity for economic, recreational, cultural and ecological purposes. Not only do we take many primary materials and draw great benefit from it, but they also regulate the biosphere. The values of the earth's biological resources can broardly classified into

A. *Direct economic values* – Consumptive uses and productive uses.

B. *Indirect economic values* – Non-consumptive uses and options for the future.

A. Direct Economic Values

One of the most fundamental values of the biological resources, especially plants, is in providing the world's food. Very few of the species with potential economic importance actually reach world markets. Perhaps 30,000 species of plants have edible parts, and throughout history a total of 7,000 kinds have been grown or collected as food. Of the latter, 20 species provide 90% of the world's food and just three wheat, maize and rice – supply more than half (Wilson, 1992). Besides food, biological resources provide us with all kinds of products; fibers for clothing, building materials, colourings, synthetic substances, medicines and so on. They are the basis for most human initiatives and a significant proportion of the global economy depends on them. Microorganisms, which play a vital role in many ecosystems, have enabled considerable progress in the manufacture of edible products.

Many animal or plant species have proved to be very useful for medical purposes. Through millions of years of evolution, they have evolved many special chemicals to meet their special needs in their respective niches, which can be exploited by man. The neem tree (*Azadirachta indica*), a native of India and Tropical Asia, is known as 'village pharmacy' (Wilson), because of its medicinal value, besides its insecticidal properties. The saliva of leech contains an anticoagulant called *hirudin,* which is used to treat haemorrhoids, rheumatism and thrombosis. The saliva of the vampire bat of Central and South America contains a substance to prevent heart attacks. The Malayan pit viper contains a substance called *ristrin* having similar properties. Caribbean sponges can be used to tackle rejection of organ transplants. *Rauvolfia serpentina,* the serpent wood plant has been traditionally used in India to treat snake bite, nervous disorders, dysentery, cholera and fever. *Reserpine* an extract from the plant, has become the principal source of materials for tranquilizers. Domesticated animals have given us hormones and enzymes, while fungi and microbes provide life saving drugs such as antibiotics such as penicillin and tetracycline. Biological resources have immense potential for being ingredients of medicines to cure several diseases. Fireflies produce a flash of light from the tip of their abdomen, the reaction being formed by the luciferin-luciferase complex, producing bright twinkling light, using little heat or energy. Scientists study this insect to explore possibilities of producing a cool, inexpensive and energy-efficient light.

There are several bioresources which are products of rich biodiversity of our forests. We use timber in some form or other as fuel wood or in house construction or as furniture. The paper that we use also comes mostly from the trees. The non-timber forest products like honey, bee wax, fibers, wild meat, skin, wild fruits, vegetables, plant gums, resins and medicinal plants are equally valuable as that of traditional timber products. The biological resources also form the primary sources for industrial products. Another direct value of the biodiversity is in the form of ornamental plants and orchids which have high market value nationally and internationally.

If we have to increase productivity and biological wealth with a minimal disturbance of natural ecosystems and loss of biological diversity, instead of relying solely on a handful of cultivated crops and domesticated animals, we should explore the possibility of raising a few wild animals and plants commercially for food products. For example, the giant Amazon river turtle of the genus *Podocnemis*, is an easily cultured species on a commercial basis that is far superior to cattle as a potential source of high protein meat. This tends to save the nearby forest land from the depredations of overgrazing by cattle. Present day aquaculture also utilizes only a small fraction of available diversity. Almost 300 species of fish are cultured for food in the world. But 85% of the yield comes from only several species of carps, while tilapias contribute a larger part of the remainder of the 18,000 species of fish known to science. Presently, at least 10% of them can be exploited commercially which will vastly increase the utilized diversity (Wilson, 1992). In this context it is wise to exploit microcrustaceans like *Daphnia* (water flea) as fish feed. They will generate almost 20 metric tons of fish flesh per hectare within five weeks. This is ten times the production rate for legumes like soyabeans and at one tenth the cost per unit of protein produced (Myers, 1997).

Conservation biologists and economists are developing a new method, called *ecological economics* to assign monetary value to biological diversity. This provides economic justification for conservation of trend diversity. A number of methods have been developed to assign economic value to biological diversity. In one method, resources are divided between *direct use values* which are assigned to products harvested by people and *indirect use values*, which are assigned to benefits provided by biological diversity that do not involve harvesting or destroying the resources. *Option value* is assigned to the potential future value of biological diversity; and *existence value* is based on the willingness of society to pay for the protection of biological diversity and to allow contribution of ecological and evolutionary resources without disruption. Direct values can be further divided into *consumptive use value* and *productive use value*. Consumptive use value is assigned to products that are consumed locally, such as fuel wood, wild meat, fruits and vegetables, medicinal plants and building materials. These goods can be valued by determining how much money people would have to pay for them if they were unavailable in the wild. Productive use value is assigned to products harvested in the wild and sold in the markets such as commercial timber, fish and shellfish and wild meat (Primack, 2002).

B. Indirect Economic Values

The components of biological diversity that provide economic benefits without being harvested and destroyed during use are said to have indirect use value.

Ecosystem integrity and health values

Biological integrity is an ecosystem's ability to maintain a balanced integrated adaptive community of organisms having a species composition diversity and functional organization comparable to that of the natural habitat of the region. Biological health is the state in which the genetic potentials of an ecosystem and member species promote its stability and sustainability. Biodiversity provides "insurance" or a buffer against environmental fluctuations, because different species respond differently to these fluctuations, leading to more predictable collective community or ecosystems properties. Species that are functionally redundant for an ecosystem at a given time may prove to be very useful at a later time. From a functional point of view, several species may have an indirect use value, so far as their individual traits and interactions contribute to maintain the functioning and stability of ecosystems and biological cycles. In soils, the termites, earthworms and ants are *ecosystem engineers* that may have important roles in maintaining ecosystem health. As these organisms move through soils, they mix organic and mineral materials to soils, form organ mineral complexes that influence soil structure. The fundamental soil processes such as carbon mineralization, carbon sequestration and stable pools of complex organic matter, nitrogen fixation, nitrification and other processes are important determinants of plant growth. Invariably ecosystems with a lower diversity of species will be less able to adapt to the altered weather conditions associated with rising CO_2 levels and global climate change. Coral reefs and mangroves protect the adjacent land, thereby reducing the effects of sea erosion on coasts. Water purification, flood control, coastline stabilization, waste treatment, disease regulation, maintenance of air quality and biocontrol of pests/pathogens are a few important services rendered by some components of diversity. According to the oft quoted *Gaia Hypothesis* (Lovelock, 1988), the earth is a super organism and biotic processes are the major regulators of physical processes of the earth and the earth's temperature and the oceans salinity are controlled by biodiversity.

Recreational Values

Biological diversity enshrined in parks and sanctuaries especially in the tropics, features prominently in the growing recreation and ecotourism industry. Recreation rejuvenates humans. The number of people involved in nature recreation and the amount of money spent on such activities are surprisingly large. Ecotourism is a special category of recreation involving people visiting places and spending money to experience unusual biological communities such as tropical evergreen forests in Western Ghats of south India, Coral reefs near Andaman Islands and the deserts in Rajasthan and to view particular 'flagship' species such as elephants in Mudumalai, Bandipur sanctuaries, lions in Gir forest or rhinoceros in Kaziranga Sanctuary in Assam. The revenue guaranteed by ecotourism justifies protection of biological and habitat diversity especially when activities of ecotourism are integrated into overall conservation and management plans.

Educational, Intellectual and Scientific Values

Nature themes often find a place in many school books, television programes and movies. The information value of biodiversity is reflected in various aspects of biomechanics and bioengineering of which genetic engineering, the base of biotechnological applications is a facet. Deriving phylogenetic and biogeogrphic relationships between related plants and animal taxa is an intellectual feast to scientist which helps to unravel the evolutionary history of the biota, both extinct and extant.

Psychospiritual, Religious, Ethical and Intrinsic Values

Aesthetic beauty, religious awe and scientific knowledge are some examples of the psychospiritual value category. E.O. Wilson used the term "*Biophilia*" to describe the inherent positive attitudes of love for nature and for living organisms by humans which expresses as wonder and awe when he/she ponders over nature's mystery.

Many animals and plants are overrated and conserved for religious sanctity. The Indian national bird, peacock (*Pavo cristatus*) is sacred to the worshippers while Hanuman langurs and cows are venerated. Many religions have traditions of "wandering in the wilderness" in order to communicate with God. Moses, Isaiah, Jesus and St. Francis of Assisi, the Hindu sages and saints and Sufis of Islam all sought out the solitude of wilderness. Being in the wilderness keeps the mind focussed and the meditative minds soar high to experience the transcendent. The most central ethical arrangements emphasize that humans have a duty to protect species based on their intrinsic value unrelated to human needs.

The intrinsic value of nature finds beautiful expression in the simple life styles and intrinsic damages of the proponents of environmental activism known as *deep ecology* (Naess, 1989; Sessiour, 1995). Deep ecology is based on biocentric equality in contrast to atmospheric superiority and human happiness. It advocates that "all things in the biosphere have an equal right to live and blossom and to reach their own individual forms of unfolding". They believe in local control without bureaucratic constraints organized according to ecosystems or bioregions.

Aesthetic and Cultural Values

The beauty inherent in biodiversity is a great source of pleasure although this aesthetic value is impossible to quality its value is rather fundamental to humans, who need a varied natural environment. The aesthetic aspect of biodiversity is not simply a matter of enjoying the beauty of particular sites, but rather a genuine need which is deeply rooted in every person, as the diversity of life adds to its quality. We carry within us the desire to protect and improve the environment for present and future generation. Landscapes also reflect cultural history. We value this diversity because it contributes to our sense of belonging. It is the source of our different ways of seeing reality. Landscapes have embodied local history and inspired the imaginations of different peoples for thousands of years.

Evolutionary Forces behind Generation of Biodiversity

One of the astonishing and mysterious facts about the generation of biological diversity is that this diversity is created from so little physical matter. The biomass of all living organisms makes up only about one point in ten billion of the earth's mass. The key to the generation of biodiversity is *biological variation*. Variation develops at the level of genes, species, community and landscape contributing to *shape biological diversity* at respective levels. Variation within a species results ultimately from mutation and recombination of genes. Since certain genes confer a competitive advantage in a given environment, individuals having these genes will move successfully, survive and reproduce. In other words, natural selection which is differential reproduction of genes of adaptive value to a given environment promotes evolution. New species arise when populations diverge genetically and are no longer able to interbreed at least in the wild. Reproductive isolation can evolve in several ways and usually follows the physical isolation of populations by geographical barriers such as rivers, oceans or mountain ranges. Any physical or geographic feature that prevents movement by plants or animals can isolate populations and allow new species to evolve. Alternately, isolation can lead to extinction if the isolated populations are too small. Thus evolution is absolutely a phenomenon of populations.

Geographical or allopatric speciation is supplemented in nature by other modes of speciation of which *sympatric speciation* is very common. It is due to the sudden appearance of reproductive isolation and the origin of a new species in the same place as the parent species, in contrast to *allopatric speciation* in which new species originate in a different place while isolated by a physical barrier. Sympatric speciation through evolution of the intermediate host races of insects that feed on plants (phytophagy) is a well documented process in this category.

Even though species diversity can explode quickly and extensively by such modes, the threat of extinction both mega extinctions and ordinary extinctions are a source of botheration to conservationists. This is because of the unique advantage of naturally rich biological diversity in the different biomes. This takes long stretches of geological time and the accumulation of large reservoirs of unique genes to build the ecosystems slowly, over millions of years and this slow and steady process is unfortunately irreversible.

Great creative potential, resulting in the origin of *evolutionary novelties* is contributed by the forces of evolution namely, the vast array of *mutations*, with ability to assemble complicated new structures, and to physiological processes including new patterns of adaptive behaviour in very short periods by geological standards. Besides *genetic drift* involving randomly fixing genes irrespective of their adaptive value in small populations, the forces of *immigration* are also involved. An important version of genetic drift accelerating the favouring of new species is the *founder effect*.

Evolutionary Forces behind Maintenance of Biodiversity

Biological diversity, thus generated by the interplay of organic evolution in the geobiological history of the earth, is again sustained through such evolutionary phenomena as *adaptive radiation*, *convergence* and *ecological release*. *Adaptive radiation* is the term applied to the spread of species of common ancestry to different niches. *Evolutionary convergence* is the occupation of the same niche by products of different adaptive radiations, especially in different parts of the world. The adaptive radiation of sharks has resulted in such bizarre forms like hammer headed sharks, tiger sharks, sharks with sawed rostrum (*Pristis*) and ray-like angel sharks (Fig. 2.1). Again woodpeckers all over the world and woodpecker-like forms illustrate the *dual patterns of adaptive radiation* and *evolutionary convergence*. Another classical example of adaptive radiation is found in the freshwater fish family Cichlidae. This prolific fish family occurs from southern USA to South America in the new world and from Egypt to South Africa in the old world. An array of primitive taxa also lives in Madagascar (Malgasey) and around 3-4 species are endemic to Southern India and Sri Lanka. *Ecological release* is a phenomenon, when a single species occupies multiple habitats due to scarcity of competitors.

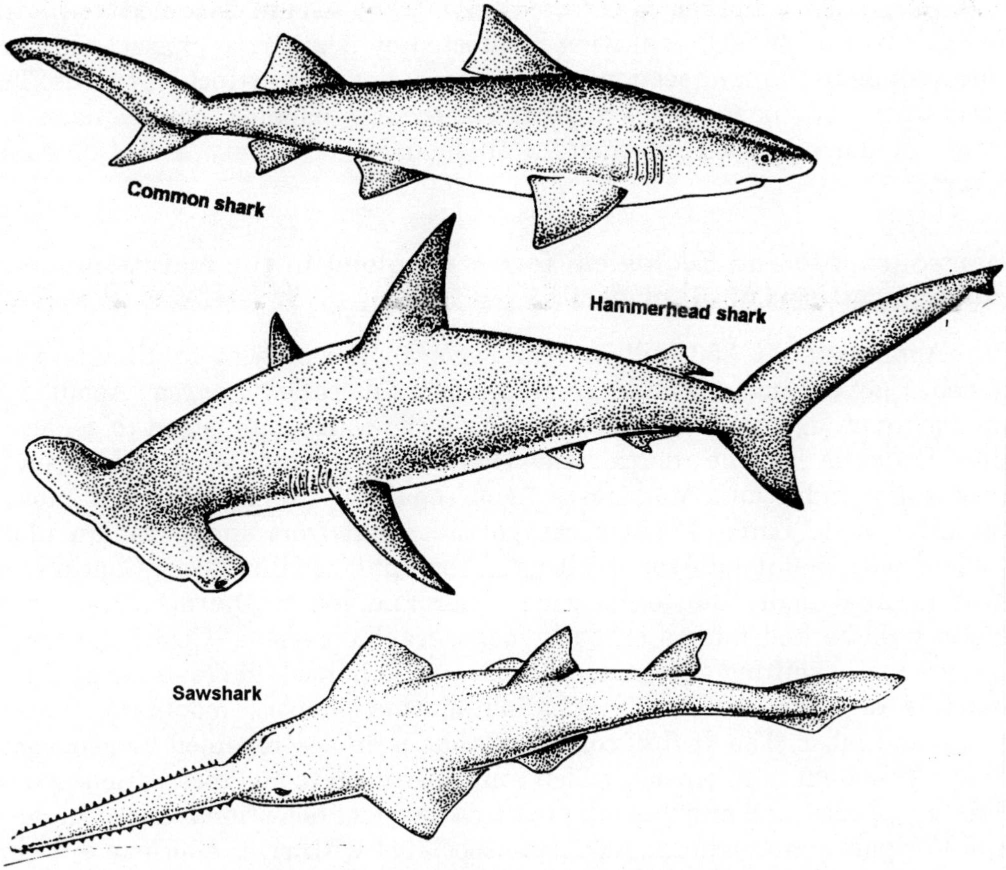

Figure 2.1. Adaptive diversity of sharks

Generation and Maintenance of Biodiversity on Islands

Islands are of two types, viz., continental islands, which are close to the main land (e.g. Sri Lanka) and oceanic islands, which are remote islands in mid ocean which have volcanic origins (e.g. Galapagos). Islands are inherently surrounded by barriers and many islands have plants and animals that differ drastically from the mainland areas with which they are associated. The species-area relationship was first noted with respect to islands. It was found that as the area of any habitat declines, so does the number of species. Later studies extended the species-area effect to habitat islands such as caves, isolated wetlands, grasslands and forest fragments. As area increases, so does the diversity of physical habitats and resources, which in turn support a large number of species.

MacArthur and Wilson (1963, 1967) proposed the equilibrium theory of island biogeography according to which, the number of species on an island represents a balance between immigration (or colonization) and extinction. In this model, colonization is affected mainly by island distance from the mainland (near or far); extinction is affected by island size. Species richness corresponds to the intersection of the colonization and extinction curves. The greatest number of species is predicted to occur on islands that are near and large. A more detailed information on Island Biodiversity is presented in Chapter 14.

Biogeographic and Ecological forces involved in the maintenance of current patterns of biodiversity

Approximately 250 million years ago, all the present continents were joined together into a single great continental, called *Pangea.* About 200 million years ago, in the cretaceous period the continents began to separate into Laurasia in the north and Gondwana in the south. Eventually peninsular India and Madagascar got separated from Antarctica moved towards north. Later, Madagascar got detached from India. Indian plate collided with mainland Asia resulting in the uplift of Himalayas. Continental drift isolated many life forms each other and led to the development of distinct floras and faunas on each biogeographic regions. Geological forces related to the shifting of continental plates also helped, to create a variety of habitats on each continent, including marine shelves, mountain ranges, rivers and lakes. The drift of continents has been accompanied by geological uplift. This uplift and erosion which continues even today, created many new habitats. Plants and animals adapted to a range of elevational microclimates and to riparian and aquatic habitats associated with river and lake systems are well known.

Tectonic activities also shape the spatial patchiness in the distribution of global biological diversity. There is a strong spatial association between tectonic activities and the areas of high biological diversity, especially in the tropics. Tectonic activities over long geological time periods bring about altitude variations in their surrounding areas, contributing to volcanic mineral nutrients and bring about climatic changes. All of the above phenomena promote habitat heterogeneity, facilitating a high species packing in such area.

Besides, these natural catastrophes like meteorite strike, glaciation events (periodic advances and retreats of continental glaciers since Pleistocene) have created new expressions of biodiversity. Ecological processes which are crucial to shape and maintain biodiversity, include energy flows through the ecosystems, nutrient cycles, hydrological systems and natural disturbance regimes like fires, windstorms, landslides and floods. Any change in the historical disturbance regime of an ecosystem may alter species composition by reducing the importance for invasive species or both.

Biodiversity: Issues and Priorities

The need to maximize the contribution of biodiversity to human welfare particularly in the developing countries, coupled with minimizing the cost of biodiversity loss is an important aspect which has to be prioritized. In this effort the values of biodiversity important for developing countries, how much of biodiversity should be conserved and the relative priorities in terms of species, genes and habitats are significant aspects. The primary effort naturally will relate to the value of biodiverse livelihoods and the genetic improvements of locally important crops. Consideration should also be made to the potential value of unknown species for industrial or medicinal values. The productive use-value of biodiversity in the form of genetic resources for developing agriculture in a less productive environment, besides the priority accounted to Tropical moist forests are also to be considered, though with lesser priority. While the International Union for the Conservation of Natural Resources (IUCN) has suggested *distinctiveness*, *threat* and *utility* as thumbrules, the overall aspects of significance also include their value to developing countries, gaps in the programmes and aesthetic values may also be a factor in determining priorities. However the choice of priority genes/species/ecosystems, must be guided by some criteria relating to the socio-economic values of different biological resources. Priority should be accounted to these habitats with the greatest number of species or to the 'Centres of diversity'. The *degree of endemism*, *ecological stress* and *distinctiveness* should be added considerations, besides threats to biodiversity due to diverse anthropogenic pressures. Decisions have to be made on the relative merits of each criteria to determine the strategic priorities. For example, Tropical moist forests are the clear priority, if species richness in terrestrial ecosystems is the prime criteria.

Priorities at the species and habitat levels will be more difficult to determine in setting species priorities. Distinction has to be made between domesticated species, wild relatives of domesticated species, species which might be useful and species which are used, but not domesticated. Needless to emphasize that rigorous priorities and implementation of cost effective conservation strategies are important considerations (ODA, 1991).

CHAPTER - 3

PATTERNS OF BIODIVERSITY

The distribution of taxonomic diversity on earth is spatially heterogeneous. This encompasses the number of species and numbers of supra-specific taxa (genera, families, orders, etc.) distributed globally. Viewed from the temporal perspective, the origins of contemporary biodiversity are on the evolutionary time scale of speciation and extinction. Patterns of biodiversity and factors responsible for the splitting and extinction of plant and animal taxa are analyzed in the context of evolution of living organisms over time.

Global Pattern of Biodiversity

A hierarchy of factors is responsible for the accommodation of an enormous number of species on earth. Geological events like continental drift and plate tectonics have led to present separation of the earth into five continents and some what parallel diversifications of each facilitate the evolution of a large number of species. Wide dispersion of landmasses gives rise to a large number of provinces. The dispersed land masses allow independent evolution of terrestrial and freshwater biota, with simultaneous independent evolution of marine biota in the oceans within a continent. The large-scale gradients of environmental variables and resources, cause each species to have optimal performance on a particular point; or these gradients could result in a massive diversity. However, diversity is unevenly distributed on the earth.

Patterns of distribution of diversity on earth are also very much scale dependent. Biogeographical processes play a major role in determining diversity patterns at the meso- and macro scales. Broad survey of geographic patterns of diversity at the global level reveals that many patterns in taxonomic diversity remain essentially similar. This is irrespective of whether diversity is quantified in terms of correlations of species richness across space or correlations of species richness and taxonomically based measures of character diversity. Furthermore, differences in the sizes of areas (i.e. area effect) usually, though not invariably, influence the spatial patterns of taxonomic diversity.

Spatial Patterns of Taxonomic Diversity

Gradients in diversity

A cline of diversity can be seen from the equator to the poles, with highest diversity in tropical regions and progressively lowered diversity at higher latitudes for many groups such as trees, mammals, birds, reptiles, amphibians and insects. In other groups no such correlation is evident, or there are increases to a maximum in temperate zones (e.g. freshwater plankton, bees, ichneumonid wasps, psyllids, aphids and sawflies).

Latitudinal gradients in taxonomic diversity may be explained as a result of a large number of possible mechanisms. These include the effects of competition, mutualism, predation, patchiness, and environmental stability, productivity, area, number of habitats, ecological time, evolutionary time and solar energy. 'Greater species diversity is due to greater' "effective" evolutionary time (evolutionary speed) in the tropics, probably as a result of shorter generation times, faster mutation rates and faster selection at greater temperatures' (Reid, 1992). Rosenzweig (1995) is of view that "latitudinal gradients arise because the tropics cover more area than any other zone". Their greater area stimulates speciation and inhibits 'extinction'.

Diversity patterns across Biological realms

A comparison of faunal richness between marine and terrestrial realms reveals that relative richness varies at different taxonomic levels. For example, whereas 32 phyla of animals are represented in marine systems in contrast to only 12 phyla on land, just 15% of currently named species are known from marine waters, despite the vastly greater area occupied by the oceans. May (1994) gives five reasons to explain the preponderance of animal species on land compared to the sea: (i) Life originated in the sea and early metazoan diversification was predominantly marine contributing to more numerous high-ranking taxa (families orders and phyla) in the oceans compared to land; (ii) continental environments are more heterogeneous than marine one; (iii) The ocean bed environment is less "architecturally elaborate" than the terrestrial environment; (iv) Patterns of herbivory differ between land and sea distinctions of marine and terrestrial species assemblages.

Diversity patterns across Biogeographic regions

The classical biogeographical regions are the Palaearctic, Nearctic, Neotropical, Ethiopian, Oriental and Australian. Of the six biogeograpical regions commonly delimited in the terrestrial realm, the Neotropics is generally considered to comprise the greatest overall levels of taxonomic diversity (around 50% of total number of species distributed globally). Tropical species richness descends from Neotropics to the Indotropics to the Afrotropics with regard to insects (Gaston and Hudson, 1994). Mammalian diversity is at its peak in the Afrotropical region. There are 52 families, 298 genera and 1045 species of mammals in the Afrotropical region, compared with 50 families, 309 genera and 1096 species in the Neotropics and 50 families, 312 genera and 995 species in the oriental region (Cole *et al.*, 1994). The pattern varies even within various orders of insects. For instance, swallowtail butterflies (Papilionidae) are far more speciose in the Indotropics (Indotropics 225 species, Neotropics 116 species, Afrotropics 53 species), Dragonflies have similar levels of diversity in the Indotropics and Neotropics (Indotropics 1435 species, Neotropics 1473 species, Afrotropics 881 species). Marine diversity is highest around coral reefs. The Indo-West Pacific is the richest region both for species of reef fish and for genera of corals (Gaston and Williams, 1993).

Diversity patterns of endemic taxa

A taxon is endemic to an area if it occurs there and nowhere else. Endemism is an extremely important factor in a species risk of extinction. For example, if the populations of an endemic species in Madagascar, which harbours high levels of endemism for many groups of organisms, go extinct the species will be globally extinct. On the other hand species on mainlands often have many populations distributed over a wide area, so that the loss of one population is not ruinous for the species as whole. Species that occupy a small area because they have only recently evolved from closely related species of cichlid fish that occupy Lake Victoria come under this category. In contrast, paleoendemics are ancient species whose close relatives have all gone extinct. Examples include the giant panda and the coelacanth fish, a relic of an array of fishes that flourished before the Devonian age of the Palaeozoic Era. The phenomenon of microendemism as seen in the distribution of species of mayflies, caddis flies and fresh water fishes in Madagascar, wherein the occurrence of a single species is restricted to a river basin or small groups of basins in the eastern rain forest region, is also equally important from the conservation point of view.

Centres of endemism are areas where many narrow endemic species occur together. They range in size from small strips to the entire biomes.

Endemism correlates generally but not completely with species richness. The great numbers of endemics are found on islands or in isolated mountain peaks, deserts springs and other unusual localised habitats where prolonged geographical isolation of populations has promoted rapid speciation.

Distribution of phylogenetically based species and the 'hotspot' concept

Number of species alone is an inadequate guide to the relative importance of areas. Any robust measure of biodiversity must also take into consideration the phylogenetic relationships of the species involved. If our goal is to maximize the diversity that is to be conserved, we should lay more emphasis on the conservation of the species belonging to a tiny evolutionary lineage, that represents the sister groups of a much larger lineage rather than an individual species belonging to that much larger lineage. For instance, the discontinuously distributed lung fishes which are the uncles of Amphibia are to be given higher priority rather than a species taxon. By this standard some isolated south temperate areas of endemism and hotspot island continents like Madagascar seem to deserve conservation priorities rather than over emphasizing the whole tropical belt for conservation.

The concept of 'hotspots' a term borrowed from the physical sciences and popularized by Myers (1988), refers to areas with high species richness and high concentrations of endemic species which are experiencing exceptional threats of habitat modification or loss. This is presently widely used in conservation planning. Myers initially identified 10 areas comprising 3.5% of remaining 'primary' tropical forests occupying only 0.2% of the land surface of the earth but housing an estimated 13.85% of the world's plant species. He later included 8 additional areas, 4 in the tropical forests and 4 in areas of Mediterranean type vegetation. One of the consequences of the initial 'hotspot' analysis and the identification of tropical megadiversity countries (such as Equador, India, Indonesia, Madagascar) was the relative neglect of biodiversity in temperate regions, especially the warm temperate or Mediterranean regions of the world (Heywood, 1995).

Myers *et al.* (2000) subsequently increased the hotspot areas to 25, based on species endemism, and degree of threat. To qualify as a hotspot an area must contain at least 0.5% or 1,500 of the world's 300,000 plants species as endemics. The four vertebrate groups, mammals, birds, reptiles and amphibians, comprise 27,298 species consisting of 4,809 mammals, 9,881 birds, 728 reptiles and 4,780 amphibians. Due to priority of data, fishes and invertebrates were excluded from the analysis. Vertebrates were used as a back-up support and also to determine congruence and to make other comparisons among the hotspots. Myers (2000) justified the selection of 25 global 'hotspots' due to the observation that as many as 44% of all species of vascular plants and 35% of all species of amphibians, reptiles, birds and mammals are confined to 25 hotspots comprising only 1.4% of the land surface of the earth. They are very much threatened.

Table 1. Comparison of twenty-five hotspots

	Hotspots	Plants	Number of Birds	Mammals
1	Tropical Andes	45,000	1,666	414
2	Central Chile	3,429	198	56
3	Choco/ Darien/ Western Ecuador	9,000	830	235
4	Mesoamerica	24,000	1,193	521
5	California region	4,426	341	145
6	Caribbean	12,000	668	164
7	Brazilian Cerrado	10,000	837	161
8	Atlantic forest of Brazil	20,000	620	261
9	Guinean Forests of west Africa	9,000	514	551
10	South African Karoo	4,849	269	78
11	Cape region of South Africa	8,200	288	127
12	Eastern Mountain forests of Kenya and Tanzania	4,000	585	183
13	Madagascar and Indian Ocean islands	12,000	359	112
14	Mediterranean Basin	25,000	345	184
15	Caucasus region east of the Black sea	6,300	389	152
16	Western Ghats and Sri Lanka	4,780	528	140
17.	Indo-Burma	13,500	1,170	329
18	Mountains of south central China	12,000	686	300
19	Sundaland Island region	25,000	815	328
20	Wallacea Island region	10,000	697	201
21	Philippines	7,620	556	201
22	Polynesia/Micronesia	6,557	254	16
23	New Caledonia	3,332	116	9
24	New Zealand	2,300	149	3
25	Southwest Australia	5,469	181	54

Source: From Mittermier *et al.*, 1999.

Longitudinal Gradients

In contrast to latitudinal gradients in taxonomic diversity longitudinal patterns have been largely ignored. However broad gradients in taxonomic diversity with changing longitude have been for a few taxa reported. These include increases in insect and mammalian species diversity towards the west of North America and an increase in water bird species richness towards the East of Southern Africa.

Altitudinal and Depth Gradients

Altitude invariably exerts an influence on distribution of taxonomic diversity, though not to the extent of latitude. The higher the altitude the lower the temperature. Spatial measures of taxonomic diversity should always take into account the extent of landmass in terrestrial areas. Two broad patterns have been documented, a decline in species richness from low to high altitudes and a peak of richness at mid-altitudes. While considering altitudinal gradients of distribution of diversity, ecological interactions, latitude, disturbance and sampling frequency should also be taken into consideration. In many other taxa, alternate pattern of simple declines in taxonomic diversity with altitude have also been observed. At local and regional scales, the topographic diversity of the habitat also is a major determinant of diversity.

In some respects, depth can be regarded as the marine equivalent of altitude. In the pelagic and benthic realms, diversity usually peaks at intermediate depths. Species richness is at its peak in 1000-1500 m depth range for pelagic assemblages. Maximum diversity is noticed at shallower depths for marine animals higher tropic levels of the food chain. The vertical depth range for deep-sea benthic species is highest at intermediate depths and decreases towards either shallow or abyssal boundaries (Gaston and Williams, 1993).

Ecological analysis of biodiversity

For purposes of comparing communities and ecosystems some of the prevailing systems of analysis relate to: (1) *Species-area relations* wherein the number of species encountered is proportional to a power of the area sampled. i.e. s+CAz, where S is the number of species encountered, A, the area sampled and z is the empirical constant, C and z being diversity measures.

(2) *Alpha (α) diversity* which is the species diversity within a community or habitat, comprising species richness and evenness.

(3) *Beta (β) diversity* is the intercommunity diversity expressing the rate of species turnover per unit change in habitat

(4) *Gamma (γ) diversity* is the overall diversity at the landscape level, which includes both α and β diversities. It is expressed as $\gamma = \alpha \times \beta \times \theta$, where α is the average value of β diversity and θ is the total number of habitats or communities in the landscape unit (Schluter and Ricklef, 1993; Singh and Khurana, 2002)

Biodiversity Crisis

The most striking feature of earth is the existence of life and the most striking of life is its diversity. This biodiversity has been awe-inspiring to humans from time immemorial and has been a scientific curiosity over a few centuries, but is increasingly becoming a source of concern, even of alarm, in the past few decades due to its rapid erosion and depletion. The earth and its biodiversity are dynamic and ever changing. The diversity of species found on the earth has been increasing since life originated. This increase has not been steady. There were periods of high rates of speciation interspersed with periods of 'evolutionary stagnation' and episodes of 'mass extinction' as revealed by the fossil second. This slow continental pace of extinction and replacement has gone hand in hand with an ongoing, overall increase in diversity. The phenomena of migration and adaptation along with the emergence of new species, has generally counter balanced the effect of extinction over very long periods.

The evolutionary history of marine animals is better investigated than terrestrial organisms because their hard parts are well fossilized in rocks formed from marine sediments. Marine animals first originated about 600 million years during the Palaeozoic era. According to the paleontological evidence, new families of marine animals appeared in rapid and steady succession during the next 150 million years. For the 200 millions years that followed, the number of families was more or less constant around 400. For the least 250 million years of the Mesozoic and Caenozoic era, the diversity of families has steadily increased to its present number of over 700 families. The fossil record of marine fauna demonstrates the slow pace of evolution with new families appearing at a rate roughly one per million years (Primack, 2002).

There have been five episodes of natural mass extinction in the fossil record, occurring at intervals ranging from 60–55 million years. Fifty percent of animal families including many trilobites vanished 500 millions years ago during the *first* recorded mass extinction one, which occurred during Ordovician period. The *second* one occurred during Devonian, 345 million years ago when 30% of animal families including agnathan and placoderm fishes and many trilobites became extinct. The most massive of mega extinctions occurred 245 million years ago at the boundary of the Permian and Triassic periods of geological history. At that time the *third* mass extinction with as many as 96 per cent of species of marine animals, 50 percent of animal families, many trees, amphibians, most bryozoans, brachiopods and all trilobites were wiped out. 180 million years ago the triassic mega extinction (*Fourth* mega extinction) resulted in the loss of 35% of animal families including many reptiles and marine molluscs. The latest of these episodes, the *fifth* mega extinction dates back to 65 million years, when the dinosaurs along with many foraminiferous and molluscs were wiped out. These dramatic events in the geological history of the world were due to some catastrophies like a hit by an enormous asteroid or a major volcanic eruption.

Apart from such major extinction episodes, background extinction rates are estimated as being around one species of mammal every 400 years or one species of bird every 200 years. We are presently in the midst of a *sixth* extinction episode, this one *caused by human activities rather than by a natural disaster.* With humans dominating the scene, the rate of extinction of species has far exceeded the rate of origin of new species. Over the last three centuries, 115 species of birds and 58 species of mammals are known to have become extinct. The current rates of extinction are 1000–10000 times higher than the background rates of 7–10 species per year estimated from fossil data of extinction rates is nothing compared to the loss of diversity likely to take place in the immediate future. Human induced perturbations in fragile ecosystems may induce loss of around one to ten percent of all living species over the next decade and this does not include multitudes of local extinctions, periodically occurring all over the earth.

We are witnessing the tragedy of extinction occurring almost every day. The recent Tsunami disaster is an immediate instance which has resulted in considerable loss of biodiversity. So while one lot of species is vanishing, many lots are threatened with extinction in the foreseeable future. According to the IUCN Red Data Book, the following is a broad list of threatened animals.

Mammals	507 species
Birds	1029
Amphibians	169
Fish	57
Insects	13
Molluscs	409
Corals and Sponges	154
Annelid worms	139
Crustaceans	126

The situation as regards plants is worse. Over 60,000 species may become extinct by the middle of the present century.

Cause of the Loss of Biodiversity

The proximate causes of biodiversity loss are:

A. Habitat destruction, fragmentation and degradation

The causes for the loss of species are complex and varied. But undoubtedly, the biggest reason is modification, degradation and loss of habitats.

Habitat loss threatens a wide range of taxa from plans and invertebrates to vertebrates. Habitat loss includes habitat destruction as well as habitat damage associated with pollution and habitat fragmentation. Habitat fragmentation is a process by which habitats are increasingly subdivided into smaller units resulting in their increased insularity as well as losses of total habitat area.

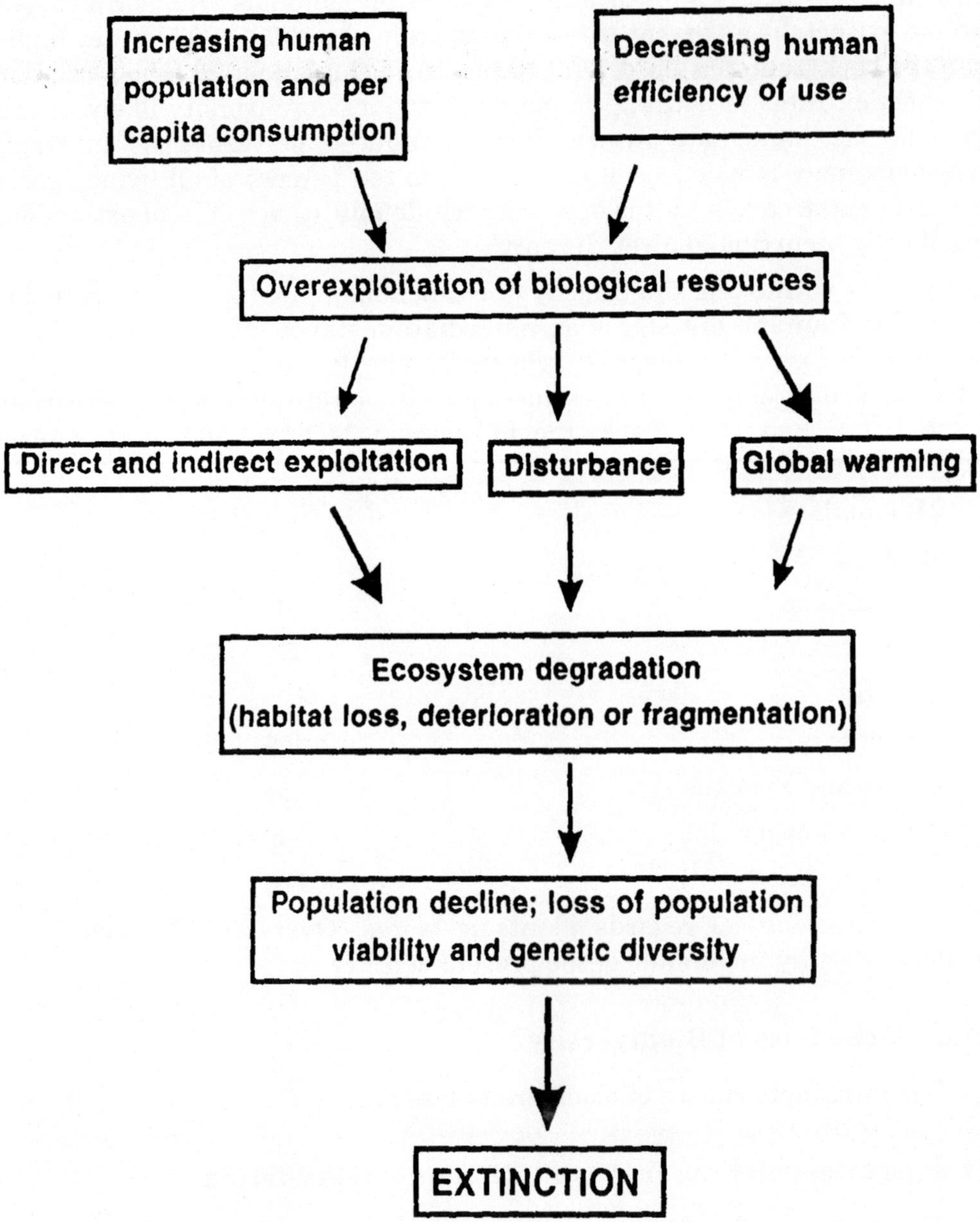

Figure 3.1. Impact of habitat fragmentation

Many animals require large habitats or, in the case of small divided ecosystems, separate habitats linked by corridors, which are routes allowing movement of individuals or groups of animals from one region or place to another. Significant fragmentation into isolated habitats can affect the breeding patterns of such animals and their search for food and shelter, resulting in gradual depletion of their genetic diversity. Habitat fragmentation also changes the microenvironment at the fragment edge. Some of the more important edge effects include microclimatic changes in light, temperature, wind, humidity and incidence of fire. Each of these edge effects can have a significant impact on the vitality and composition of the species in the fragment (Fig. 3.1). Habitat fragmentation can also be a problem for plants, preventing seed and pollen dispersal and consequently preventing reproduction. Not much is known regarding the varied interaction between species and the specific roles played by them in the functioning of ecosystems and whose disappearance would result in major upheaval. In a number of cases, several species perform the same function, thereby protecting ecosystems from premature destruction. However, the disappearance of specialized species encourages weeds and diseases to replace them. Selective extinction may thus lead to a world with fewer species which are adapted to their environment and indiscriminate increase of few opportunistic species.

Fragile ecosystems like the tropical rain forest, mangroves and tropical coral reefs are in jeopardy. Forest areas are cleared and colonized for settlement or agriculture expansion commercial logging, large hydel schemes, fire, human and livestock pressures etc. These not only decrease the overall populations of many species, but also break up their distribution patterns and render them into smaller isolated sub-populations which are more vulnerable to extinction. Drastic environmental changes can also threaten small populations, especially in fragmented landscapes, chance variation in age and sex ratios and other population parameters, termed *demographic stochasticity* can change birth and death rates. In small populations, extinction occurs quality when death outpaces births. Some species have a threshold density or numbers of individuals below which the population cannot recover. This is likely to occur when organisms modify their environment physically or chemically to encourage their survival, when group defense against predators is necessary, or when social interactions and mating successes depend upon some critical population diversity. Small isolated populations or habitat islands are prone to two kinds of genetic problems: Inbreeding depression (caused by mating between close relations) and *genetic drift* (a random change in gene frequencies). Populations with survivorship in the long run will be less able to respond adaptively to environmental change. Such a genetic erosion is particularly serious for species of current economic value–such as crop and pasture species where genetic variation is important for breeding new strains and varieties.

B. Overexploitation

Biodiversity is critically imperilled due to overexploitation, mainly for commercial purposes meat, fur, birds, hides, body organs and animals and plants of medicinal value. This is a more selective threat and particularly affects carnivores, ungulates, primates, sea turtles, crocodiles, snakes, beautiful tropical birds and timber species. Many charismatic beautiful, gastropods and bivalves for their beautiful shells, orchids and insectivorous plants are threatened by collectors and because of commercial interests.

C. Human demographic change

Demographic change in the form of population growth and migration with implications for agricultural conversion, promotion of monocultural landscapes and overharvesting techniques ultimately destabilising cultures and ecosystems. The simple growth of human population is not the whole problem. The pressure of production on biological resources is exacerbated by the unequal distribution of agricultural land and by the replacement of traditional common property rights by open access regimes. Population growth is nevertheless a powerful force for biotic improvement.

D. Introduction of exotic species

Accidental or deliberate introduction of exotic species can threaten native flora and fauna directly by predation or by competition and also indirectly by altering the natural habitat or introducing pathogens alien to that habitat. For instance, the introduction of exotic freshwater fishes like *Orochromis mossambica* and *Cyprinus carpio* have no doubt contributed much to the country's fish production, but they have also had a negative impact on the indigenous fishes in the form of competition, predation, environmental modification and so on. The unprecedented spread and colonization of the hardy *O. mossambica* has been posing severe threat to our indigenous fishes competing with the latter for space and food and quickly replacing them. The exotic common carp, *C. carpis*, a popular and angling species, causes harm to other species by stirring up bottom sediments and creating turbid conditions in impoundments. The introduction of the common carp has already eliminated the indigenous schizothoracine fishes from the Kashmir valley, *Osteobrama belangari* from the Lohtak Lake in Manipur, and the large indigenous *H. periyarensis* from the periyar lake, Kerala. Likewise, the introduction of the larvicidal fishes *Gambusia affinis* and *Poecilia reticulata* has almost replaced the indigenous *Aplocheilus parous* and *Oxzias melstigma*. Also the introduction of the gangetic carps into the peninsular rivers has affected the indigenous carps of the peninsula. These introductions generally cause *'Zoogeographical Pollution'*, loss of genetic identity of local populaions, a high level of hybridization and extinction or reduction of local communities of endemic species. Parasites and disease have also been introduced within alien species.

E. Pollution

The most subtle and universal form of environmental change causing erosion of biodiversity is pollution, commonly caused by pesticides, sewage, fertilizers from agricultural fields, industrial chemicals and wastes, and automobile and sediment deposits from eroded hillsides. Organochlorine pesticides like DDT, sprayed on crop plants to kill pests and sprayed on water bodies to kill mosquito larvae become concentrated as they go up in food chain, causing harm by a process called *biomagnification* and harm wildlife populations. Pesticides, herbicides, oil products, heavy metals (such as mercury, lead and zinc), detergent and industrial wastes directly kill organisms in aquatic environments. Human sewage, agricultural fertilizers, deterrgents and industrial wastes often release large amount of nitrate and phosphate into aquatic systems, initiating the process of eutrophication or organic enrichment, which in high concentrations harm the biota. Acid rain, ozone production and nitrogen deposition, compounds with toxic metals released into the atmosphere, green house gases like carbon dioxide and methane in the atmosphere contributing to global warming, cause great hazards in the form of air pollution to the flora and fauna.

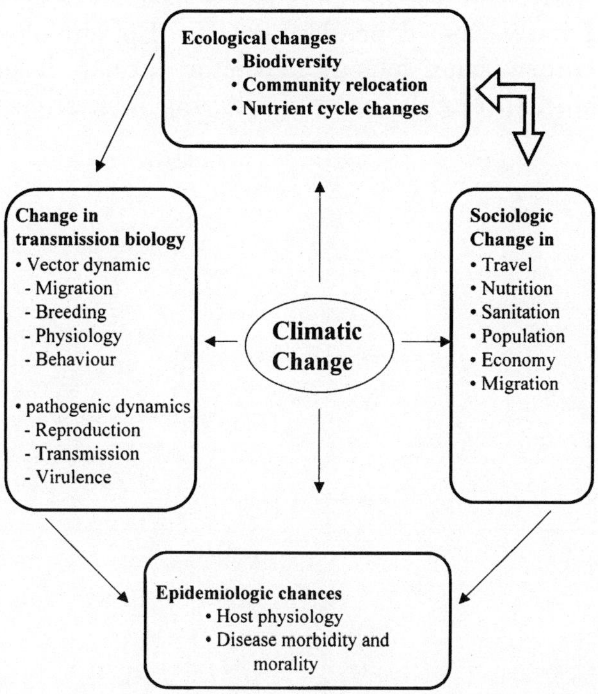

Figure 3.2. Impact of climate change (After Tyagi, 2004)

F. Global warming

A looming threat to biodiversity is global warming. The predicted warming is a result of increasing levels of carbon compounds in the atmosphere, especially carbon dioxide (CO_2) and also methane and others. Atmospheric CO_2 comes from a variety of sources, the primary ones being combination of fossil fuels (oil and coal) and the cutting and burning forests worldwide. This increase in CO_2 creates a "green house effect" that warms the earth's surface. Current estimates are that atmospheric CO_2 is increasing at a rate of 1 to 2% per year, resulting in a doubling of atmospheric CO_2 in about 50 years with resulting temperature rises of from 1 to 5°C. Sea levels would rise, patterns of rainfall would change and storms and wild fires would increase in many regions. An increased temperature of 3°C is equivalent to a latitudinal shift of about 250 km on an altitudinal shift of 500 m. Species confined to the tops of mountains will be lost due to drastic change in climate. Other species will move away from the equator unable to cope up with the change in climate. Habitat fragmentation and anthropogenic stresses along with changes in climate will have many cumulative impacts (Fig. 3.2). The other possible reasons for the loss of species could be inequality and poverty driving people to overexploit marginal lands and the breakdown of traditional mechanisms for equitable sharing and protecting resources, and economic and political policies and constraints. The earth's biodiversity being interlinked intricately, the loss of one species is most cases triggers off the consequent loss of many other dependent species. Epidemiological changes as well changes in transmission biology of vector species, besides sociological changes affect health through impact on varying intensities of vector-borne diseases.

CHAPTER - 4

FUNCTIONAL ASPECTS OF BIODIVERSITY

The study of biological diversity involves not only describing the great variety of biodiversity patterns on the earth from genes to ecosystems, but also to find out what these diverse living entities are or what their function is. Every species has a major or a minor role to perform in the functioning of ecosystems and this ecological function addresses the relationships between living entities and ecological process. 'Functional diversity' addresses the variety of these relationships. As has been mentioned earlier, Franklin (1988) and Noss (1990) clearly distinguish functional from structural aspects of biodiversity. Structure is the physical organization or patterns of a system. For example, habitat complexity is a structural measure within a community, whereas patterns of patches or corridors form important structural components in a landscape. Function involves ecological and evolutionary processes including gene flow disturbance and nutrients cycling. Huston (1994) classifies the total species found in an ecosystem into '*structural species*' and '*interstitial species*'. Structural species create or provide the physical structure of the environment e.g. reef-forming corals, sessile animals such as oysters, mussels, barnacles, tubicolous polychaetes etc. These organisms create the physical structure of their environment, produce variability in physical conditions provide resources, and in general create the habitat used by many other generally smaller interstitial organisms. Interstitial species would include most insects, other arthropods, birds, mammals and other vertebrates. 'Structural organisms' exert a major direct influence on the diversity of interstitial organisms while the latter has a minor and indirect influence on the forms through evolutionary and biographical process.

The total diversity of any community can also be broken up two distinct components viz. (1) The number of different functional types of organisms and (2) The number of functionally analogous species within each functional type. Different functional types of organisms use different resources or otherwise interact with their environment in such a way that competitive interactions are minimized. Species of the same functional types use the same resources in such a way that they could potentially compete intensely with one another.

Environmental Services of Biodiversity

Humans derive many benefits from the *environmental services* of biotas and ecosystems. Environmental services are also known as *ecosystem services*, both terms reflecting environmental functions and ecological processes. Environmental services include regulation of climate and biogeochemical cycles, hydrological functions, soil protection, crop pollution, pest control, recreation and ecotourism and a number of miscellaneous services like biological monitoring of various kinds of widespread pollution. These services are significant both from ecological and economic perspectives.

According to Myers (1996), biodiversity pays for energy and material flows, which in turn provide ecosystems with their functional properties. It supports and fosters ecosystem resilience, which is the ability of ecosystems to resist stresses and shocks to absorb disturbance and to recover from disruptive change.

Ehrlich and Ehrlich (1981) postulated a 'Rivet popper hypothesis' which states that species are like the rivets on an airplane, where each species plays a small, but significant role in the maintenance of whole ecosystems. However, if ecosystem processes are quite resistant to biodiversity decline, if they can keep on supplying environmental services after losing a good number of species and large number of populations, it is obvious that there is much redundancy built in nature. The "*Redundancy Hypothesis*" was suggested by Walker (1992), which states that many, if not most species, are not required for ordinary ecosystems functioning. Neither view is totally acceptable, nor are both the views complementary to each other. It is incorrect, then to say that each species has its essential part to play in ecosystems, let alone that it is a mainstay of stability or resilience. It is also incorrect to say that we can lose lots of species, with impunity. A cut-off stage would eventually arrive when there would be simply too few species to maintain basic ecosystem functions. Equally valid is the notion that species redundancy itself may well have a functional value as a kind of "*nature's insurance*" that becomes beneficial in the long run in essentially reducing the severe risk of collapse of ecosystems. Environmental services often depend not only on biodiversity, but also on biomass. So it is important not to confuse biodiversity with biomass and associated factors such as community make-up and vegetation structure.

Functional Diversity – functional feeding groups and feeding guilds

Functional diversity among organisms is valuable for maintaining the integrity of ecosystems and hence the services that they provide. Ecological functions are interactions of organisms with ecological processes and functional diversity is the variety of interactions with ecological processes. For example, four functional groups of terrestrial herbivorous insects are usually distinguished. They are well concealed endophytics (gall formers and stem, flower, fruit and seed borers), poorly concealed endophytics (case bearers and leaf miners), exophytics (leaf chewers) and mixed exophytics/ endophytics. Similarly, to facilitate pattern recognition in ecosystems, five functional feeding groups are broadly recognized in stream dwelling invertebrates. They are *grazers* and *scrapers* which are herbivores feeding on periphyton (i.e. attached algae and associated material); *shredders*, which are detritivores feeding on coarse particles like decomposing terrestrial leaf litter with associated microbes collectors feeding or fine (<1 mm) particles suspended in matter or deposited on the substratum; *predators* including those which swallow or engulf prey and those which pierce their animal victims extracting cell and tissues fluids, and piercers and suckers of plants cells, which are specialized microherbivores which pierce algal and other plant cells with their mouthparts and inject the contents.

Another aspect of functional diversity is the formulation of the "*guild concept*" which is a pattern of convergence. Guilds are functional groups of organisms whose members exploit environmental resources in a similar way. Species at the same trophic level that use approximately the same environmental resources are considered to be a guild of competing species. The nectar-feeding guild of birds, for example is represented by hummingbirds (Trochilidae) in the new world tropics, by sunbirds (Nectarinidae) in the African and Asian tropics, by honey eaters (Meliphagidae) in Australia and by honeycreepers (Drepanididae) in the Hawaii Islands. All are convergently similar in having a long thin beak that enables them to probe deep into flowers. Similarly, it was noted that niche segregation among cyprinid fish assemblages in stream pools of Kallar River in Western Ghats was primarily by diet. The assemblages could be divided into four feeding guilds: (1) the surface feeders (*Danio aequipinnatus, Barilius bakeri and Rasbora daviconius*) which feed largely on terrestrial arthropods; (2) a mid water generalist (*Puntius filamentosus*) which feeds on a variety of invertebrates food from bottom and water column (3) benthic feeders (*Puntius melanampyx* and *Puntius amphibious*) which feed mainly on invertebrates from different substrates and (4) benthic herbivore (*Garra mullya*) which feeds mainly on benthic algae from rocky substrate (Arunachalam *et al.*, 1997).

Food chains, Food parasitoid webs and their relationship with biodiversity

Food Chains

The transfer of energy from organism to organism sustains the ecosystem and when energy is transferred from individual to individual in a particular community, as in a pond or a lake or a river, we encounter what are called *food chains*. The number of steps in a food chain is always restricted to four or five, since the energy available decreases with each step. In other words, the energy fixed in organisms such as in aquatic plants in a pond passes through the ecosystem in the form of plant → herbivore → carnivore, where only three steps are encountered in the chain. But very often, the chains are very complicated with several steps being involved. An example may be plant → herbivore carnivore 1 → carnivore 2 → carnivore *n*. Basically two types of food chains are recognized, those commencing from the green plant through the herbivores to the carnivores (*grazing food chain*) and those starting from dead organic matter either in the forest floor or at the bottom of lakes and rivers passing through micro-organisms which break down the detritus to fundamental organic components. The detrivores are earthworm and millipedes in the forest floors and mussels and clams in the lakes and ponds (*detritus food chain*). The producer → herbivore → carnivore chain is a predator chain (Fig. 4.1). Parasitic chains also exist wherein smaller organisms consume larger ones without outright killing as in the case of the predators. Links between the grazing and detritus food chains are not uncommon and the stability of the ecosystem is directly proportional to the number of such links (Wilson and Bossert, 1971). These links are evident in situations where energy enters the detritus food chains from grazing food chains. Sometimes there occurs a re-entry of energy from the detritus food chain into grazing food chain when the organisms from the former are consumed by those in the latter. In many biological communities, one species feeds on several other species at the lower trophic level, to compete for food with several species as its own trophic level and, in turn, to be preyed upon by several species are linked together through complex feeding relationships, they form a food web. When each species involved in the food web has a number of links, variations would naturally occur in the population of each species so that the magnitude of the fluctuations of the concerned species could be predicted. If fluctuations are considerably moderate and not extremely low as to lead to extinction, then there are good chances of the species persisting in the ecosystem for a longer time.

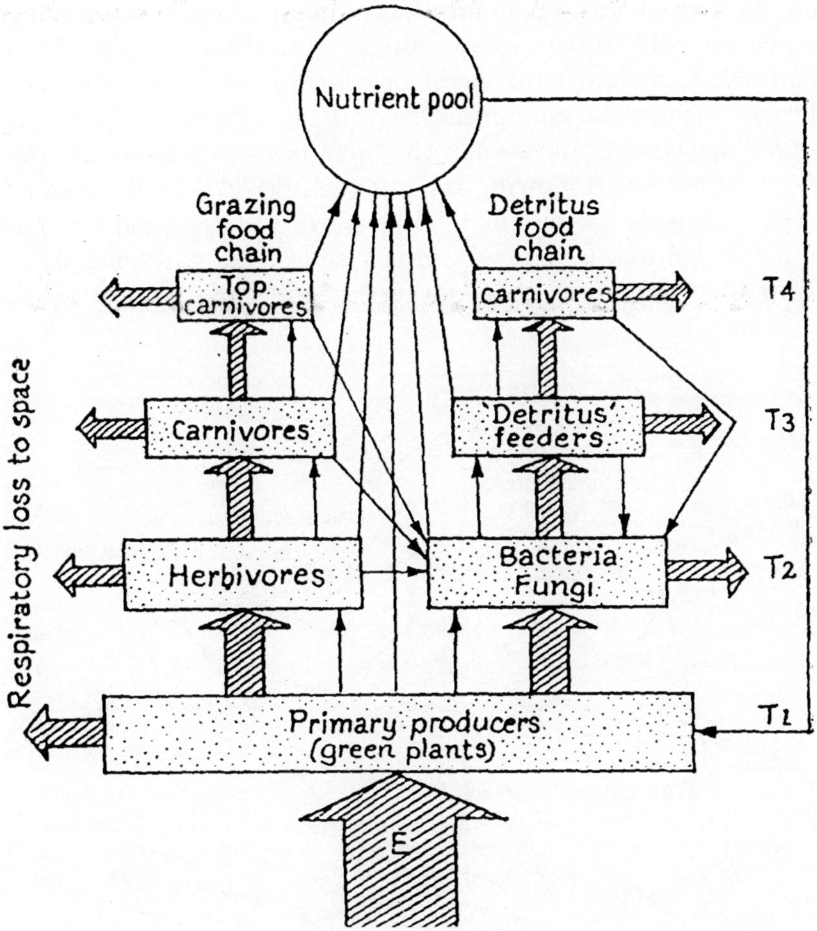

Figure 4.1. Grazing and detritus food chain

In a population of a single prey and single predator in an environmental of less complicated conditions, wide fluctuations are observed. This is often detrimental to both the species leading to their gradual extinction. When the predator depends on two or more species of prey populations, chances of extinction of both the species are diminished. This is because the availability of alternate prey would enhance the chances of the predator as well as the prey to survive. Further, predators prey upon only those species that are abundant and do not switch over to the other species until the one in handed is totally eliminated. Similarly, fluctuations are wide in prey populations living under the predation of two or more species. A wide variety of food webs are observed in ecosystems, ranging from a single food chain with one species of plant, a few herbivores, larger number of carnivores, to a more complex one with many plant species, many herbivores and larger number of carnivores (Fig. 4.2). The introduction of a predator in a lake from another source often results in the spread of its population throughout the lake, with a sudden reduction of almost all secondary consumers. These reductions in turn produce second and third order changes at other trophic levels in the ecosystem. When a dominant carnivore is removed, for example, from an exposed rocky interstitial community, significant changes occur in population density and species composition. The dominant carnivore is the starfish *Pisaster ochraceus* and following its removal, 60-80% of the available space will be occupied by *Balanus* which will soon be out crowded by *Mytilus*. This removal of the dominant carnivore in course of time will lead to conspicuous decrease in diversity from a 15 species system to a trophically simpler 8 species system.

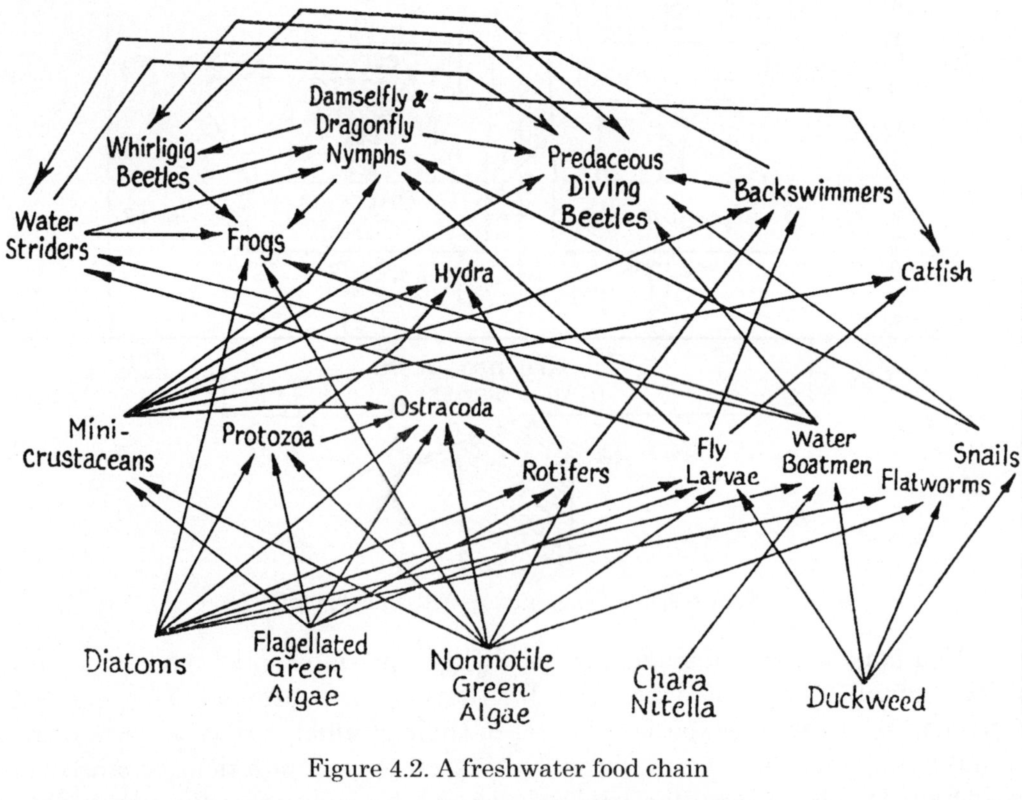

Figure 4.2. A freshwater food chain

The energy required by different trophic levels in an ecosystem is derived by the potential energy resulting from primary producers. Herbivores and some omnivores consume energy in the form of food, and these herbivores are, in turn consumed by the carnivores which have a distinctly higher efficiency of assimilation. Thus these results net secondary production which is equal to the amount of energy stored in the tissues of carnivores.

The study of the food webs is crucial to understand the ecological process underlying the patterns of diversity. Furthermore, studies of food webs containing parasitoids are essential in interpreting patterns of parasitoid diversity. Parasitoids should be studied in combination with the competitors, predators and pathogens that also influence the distribution and abundance of their hosts. Actually, the parasitoid webs are subsects of food webs that discernible the linkage between hosts and parasitoids in a single habitat. Predators and parasitoids become important for the maintenance of diversity when they regulate the populations of species which would otherwise out compete and thus eliminate other species.

Three types of food web are normally recognized *community webs, sink webs* and *source webs*. Community webs describe the relationships between all predators and prey within a habitat or set of habitats. Source webs and sink webs are focused or a single species and include, respectively, all its prey and lower trophic levels. Community webs are obviously the most valuable of the 3 types of food web as they describe a whole community and are not biased by the choice of focal species.

As a common feature of all communities mention may be made of Elton's theory of pyramid of numbers, which states that 'the animals at the base of a food chain are relatively more abundant while those at the end are relatively few in number, there being a progressive decrease between the two numbers'. In the construction of such a pyramid the first step is to group and count all autotrophs or primary producers, next is to group the herbivores or primary consumers; the carnivores are subsequently grouped into secondary, tertiary and quaternary consumers. An interesting aspect is that the pyramid is inverted when the parasitic food chains are considered. An estimation of the numbers of organisms may not give a correct picture, but their weight or biomass is taken into consideration resulting in a *pyramid of numbers* or a *pyramid of biomass* (Fig. 4.3). Both the pyramid of numbers and pyramid of biomass indicate only the number or weight of organisms in a particular ecosystem at a particular time. The term *standing crop* in this context becomes nothing but the biomass at any one time or the total amount of living matter in an ecosystem at a given trophic level. The rate of accumulation of individuals, whether it be diatoms or trees in an ecosystem, appear to be more significant in studies on energy than the standing crop. This study of the rate of accumulation of energy by individuals is called *productivity*. The term biomass accumulation ratio implies the ratio of dry weight biomass to annual net primary production. It has been shown that this ratio is very low in deserts (2-10 kg/m^2) and very high in mature forests (20-50 kg/ m^2). Both biomass and production tend to decrease with decreased moisture and here again we find that the biomass of mature forests is 20-60 kg/ m^2, while in grasslands and deserts it is 0.5-3.0 and 0.2 respectively. The importance of biomass lies in the fact that while biodiversity often plays a key role in reducing ecosystem services they can also be derived from biomass and other attributes of biotas.

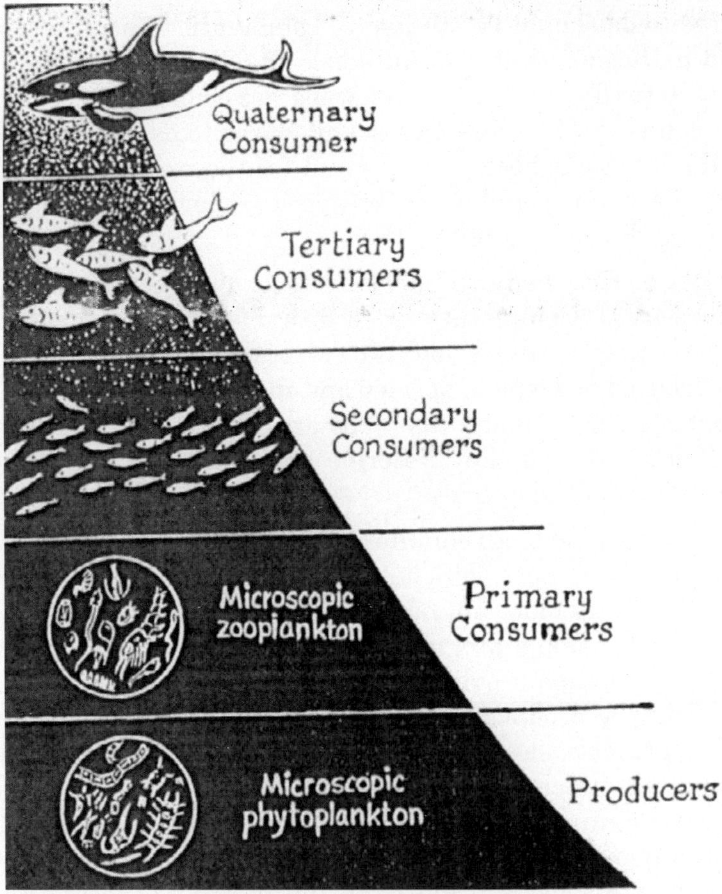

Figure 4.3. Pyramid of biomass

Species effects on ecosystem processes – The keystone species concept and ecological engineering

It is an established fact that species affect ecosystem processes at different spatial and temporal scales. However, not all species have the same influence in an ecosystem. *Keystone species* affect community or ecosystem processes to a degree that appears greater than their relative abundance in a community would indicate. In other words, species that have disproportionately large influence on the character or structure of an ecosystem are called keystone species. When keystone species are lost, removed or have their activities disrupted, numerous other species forming part of that ecosystem will also be lost and this is termed "cascade effect". Solbrig (1991) distinguished 3 main categories of keystone species.

1. Predators, parasitoids, herbivores and pathogens that contribute to the maintenance of diversity by reducing the abundance of dominant competitors and thus prevent competitive exclusion.

2. Mutualists, partner species whose fates are linked.
3. Species that provide resources which are critical to the survival of dependent populations.

Keystone predators and parasitoids

Top predators may be keystone species because predators often have a influence on herbivorous populations. The elimination of even a small number of individual predators, even though they constitute only a minute amount of the community biomass, may result in drastic changes in the vegetation and a great loss in biological diversity. For example, if top carnivores like tigers and wolves are hunted indiscriminately, deer populations will explode. The deer will severely overgraze the habitat, eliminating many herbs and shrunk species. The loss of these plants in term will be detrimental to the deer and to other herbivorous including insects also contributing to the loss of species that in habitat the soil.

Keystone resource species

Bats called "flying foxes" of the family Pteropodidae are excellent examples of keystone species. These bats are the primary pollinators and seed dispersers of many economically important tree species in the cold world tropics and pacific Islands. When bat colonies are over harvested by hunting and the trees in which the bats roost are cut down, the bat populations decline. As a result many of the tree species in the remaining forest fails to reproduce.

Parasitic Hymenoptera plays a major role in the balance of terrestrial ecosystems through their ability to regulate populations of phytophagous insects. By limiting the population size of a species that would otherwise outcompete other species, parasitoids help to maintain high species diversity. Additionally, community interactions involving parasitoids appear to be more stable than those involving predators.

Keystone mutualists

In highly co-evolved and mutually beneficial interactions the loss of one species will cause the loss of the second. Specialized pollution systems are an example of this, since it is well known that certain orchids would lose the ability for fertilization without their specific bee pollinators. In addition to such specialized systems, it has been shown that bees are essential to the fertilization and reproduction of a large number of flowering plants. The loss of bee diversity will surely be accompanied by the loss in flowering plant species. Another example of a population is found in the parasitic hymenoptera in fig wasps (Agaonidae). There are currently almost 800 species of figs (Moraceae: *Ficus*), most of which are dependent on a single, specific species of agaonid for pollination.

There are a variety of striking examples of interactions between ants and plants and many plants need their ant associates to survive or propagate. The usual benefit that plants derive from ants is protection from herbivores although other benefits include seed dispersal, pollution and supply of nutrients. In exchange for this protection plants offer ants food and shelter in the form of domatia which are structures produced by plants in the form of hollow stems, thorns or expanded leaf bases, which show no function other than sheltering ant colonies (Jolivet, 1992).

Keystone dominants

Ants influence the arthropod community in which they live in more direct ways. Ant species form mosaics which allow several species of ants to co-exist with dominant and non-dominant status. Dominant ant species not only affect and control the distribution of non-dominant ant species, but also influence both the species composition and abundance of other arthropod taxa. Since each species of ant within the mosaic can influence the structure and composition of other plant and animal species, these mosaics appear to be a significant contributory factor to arthropod diversity in the tropics.

Ecosystem Engineers

Species that extensively modify the physical environment through their activities, often termed "*ecosystem engineers*" are considered keystone species also. For example, beavers in North America build dams that flood temperate forests, creating new wetland habitat for many species. Earthworms may turn over many tons of soil per hectare each year, dramatically affecting soil fertility and thereby the plant and animal community.

Caecilian amphibians and endogeic soil reptiles which predate on earthworms, termites and ants are recognized as soil ecosystem engineers. Of significance as soil dwelling vertebrates are the caecilian species of *Icthyophis, Gegenophis, Uraeotyphlus* which play a role in subterranean predator-prey interactions (Fig. 4.5) (Measey *et al.*, 2003). Leaf cutter ants in tropical and subtropical American forests dig extensive tunnels through the soil to build fungal gardens and in the process, create new habitats for many subterranean species; their leaf-cutting activities also have a preformed effect on the vegetation.

Figure 4.4. Dung beetle diversity

Another keystone taxon that plays a significant role in the functioning of biological communities is the detritivore, dung beetle. Dung beetles exist at low density levels in tropical forests and constitute only a fraction of the biomass. Yet these beetles are crucial to the community, because they create balls of dung and bury them as food sources for their larvae. These buried materials break down rapidly, making nutrients available for plant growth. Seeds contained in the dung of fruit-eating animals are also buried, which may facilitate seed germination and the establishment of new plants. In addition, by burying and feeding on dung, the beetles kill the parasites of vertebrates contained in the dung, thus helping to keep the vertebrate populations healthy (Fig. 4.4).

Figure 4.5. Caecilian amphibians

However ecological engineers and keystone species do not appear to be the sole norm in all communities. Most communities comprise a collection of species that differ in abundance, yet all contribute in one way or another to ecosystem structure and function. Communities and ecosystems are, in fact, highly variable in space and time.

CHAPTER - 5

GENETIC DIVERSITY AND PHENOTYPIC PLASTICITY

Increased levels of genetic variation are present in most natural populations and such genetic diversity allows species to adapt over time to diverse environmental stresses. Such adaptive responses to environmental change is the result of variation among genes. Most environments are frequently changing as for instance, after a period of drought, a population will be better adapted against drought conditions. Adaptation therefore favours the organism in its struggle for existence. There are different types of adaptation, from broad adaptive zones to species specific niches. The phenotypic manifestation of genes can be modified by environmental agencies of various kinds–physical, chemical and biological. Man has relied on genetic diversity to create a wide range of genetically diverse crops as well as breeding and domesticating wild animals for desirable traits. The richness of genetic diversity in India is very impressive, particularly in the land areas of cultivated species. Agricultural and Industrial crop species offer high potential for utilization in sustainable production. Farmers have domesticated hundreds of plant species over time resulting in innumerable varieties with desirable characters as colour, flavour, resistance and so on. Opportunities provided by genetic engineering allow transfer of genes among species. For instance, a wild relative of rice in India provides a 'resistance gene' that now protects high yielding rice varieties from attack of the brown plant hopper. The unknown potential of genes, species and ecosystems represents a never ending biological frontier of inestimable value. An understanding of the total number of different genotypes present in a population (richness) and the frequency of the different genotypes (evenness) present in the same population is important. Genetic diversity will enable breeders to tailor crops to new climatic conditions. Because of the diverse array of genes, species and ecosystems, they act as resources that can be tapped as human needs and demands increase. Genetic diversity, in short, permits adaptation to a range of environmental conditions and with their change this genetic diversity confers the ability to adapt and persist. As Wilson (1992) remarks "Great biological diversity takes long stretches of geological time, accumulating large reservoirs of unique genes. By chance only a few new species are poised to move into novel adaptive zones to create something spectacular and stretch the limits of diversity". In today's scenario the potential benefits of biotechnology have to be assessed in terms of large scale genetic variation such as gene identification, regulation and expression of genes and understanding the physiological and genetic interaction between and within organisms, not to mention of the understanding of the behaviour and effects of genetically engineered organisms in the environment.

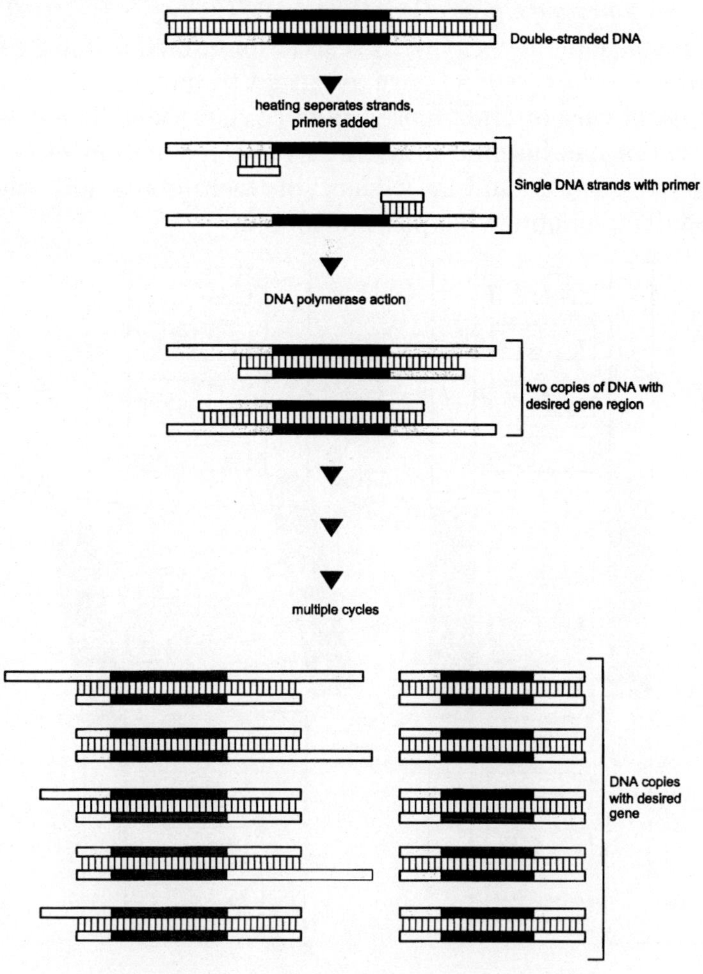

Figure 5.1. Polymerase chain reaction

Molecular Markers

In today's scenario, a gene is said to be polymorphic when it occurs in multiple forms or as multiple alleles. This can be demonstrated by gel electrophoretic studies, in terms of the banding patterns or more exactly identifying the nucleotide sequence, since the ultimate measure of genetic diversity is the nucleotide diversity, through determining the base pairs not only through a long part of the chromosome, but also in many individuals belonging to the same species. As each species comprises a vast collection of genes which are inventories of animals and plants, diversity is used as a surrogate to estimate the genetic diversity of particular species. With sophisticated DNA analysis like RFLP or restricted fragment length polymorphism, it has become possible to study the genetic characters of populations. The production of large amounts of a small segment of DNA molecule is called *amplification* and production of large number of copies of DNA is called *Polymerase Chain Reaction* or PCR (Fig. 5.1; 5.2). The products of the PCR are subsequently cut with restriction enzymes for RFLP analysis. Such molecular marker studies are of great importance in studying the genetic diversity of rare or endangered taxa (Avise, 1994).The development of molecular markers has become important for construction of linkage maps, fingerprinting of strains and a variety of techniques has been used in silkworms to detect genome-wide polymorphism.

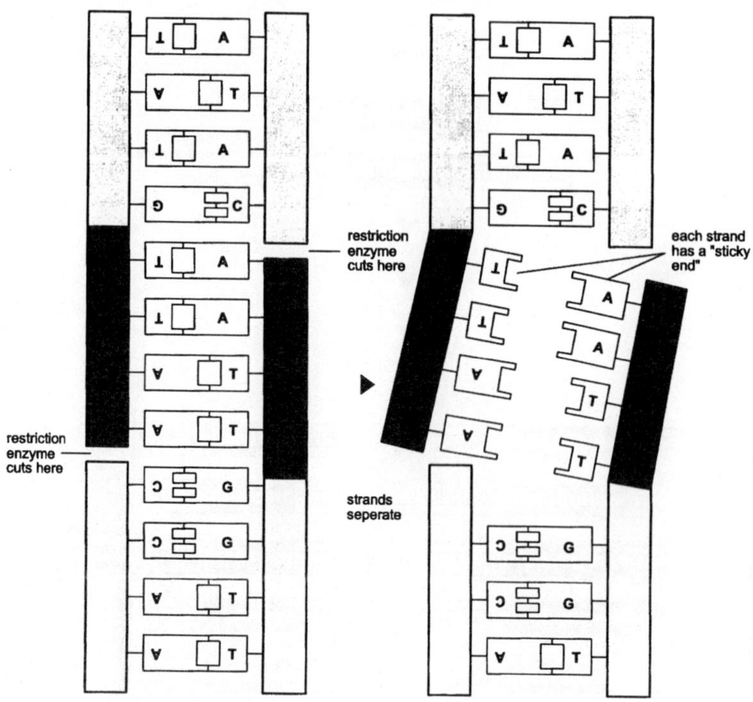

Figure 5.2. Restriction enzyme snipping

Genetic Improvement and Erosion of Genetic Base

With the formation of many breeds, people begin to ask which breed is the best. Comparisons made between different breeds show that breeds differ in respect of many performance traits and that no single breed was the best. Breeds superior in respect of some traits of economic importance are found to be inferior in others. If only one breed is found to be the best in all respects, we would certainly end up having only that breed in each class of farm animals. Similarly, studies pertaining to determining the best types from the production point of view in several class of farm animals have led to the conclusion that the relationship between type and performance was very low and that selecting for good type would not automatically lead to good performance. In short, genetic diversity is a resource for the survival of the species and future evolution (Groombridge, 1992), besides conforming fitness advantages upon the number of a population.

Subsequently, interest shifted to cross breeding followed by the eventual comparison of crossbreds with purebreds in different classes of farm animals, particularly in cattle and pigs. Such results by and large favoured the crosses, but in many cases, the differences were small. However, it was revealed that some breeds crossed better than others, and that in both crossbreeding and pure breeding, it was equally important to use good selection practices.

The normal function of the genes in the expression of various traits is quite complex, more particularly in multicellular animals. But serious efforts are underway to understand the control of gene activity as it would be of tremendous value in the future in improving the health and welfare of mankind as well as the efficient production of livestock. There are different ways in which genes can express phenotypically. In this manner, such expression can be measured objectively as well as subjectively. Phenotypic expression of genes can be divided into additive and non-additive types. The heterogenous assemblages of the immigrants species group of *Drosophila* comprises 70 nominal species which are categorized under five different subgroups. Of these the *nasuta* subgroup is a cluster of morphologically almost identical species and subspecies of *Drosophila*. The females of the different members of the *nasuta* subgroup are indistinguishable, while males have subtle differences. Cytoraces have been recognized. It is of interest, that parental races are larger in body size with reduced fertility, while the newly entered cytoraces showed reduced body size and better fitness. The occurrence of morphophenotypes is a commonplace affair and should be considered as subcomponents of a major trait enabling a holistic picture (Ranganath, 2002). But such knowledge of the different ways in which genes can express themselves phenotypically is important in order to devise effective selection and mating plans for livestock breeding.

Variation at the phenotypic level refers to the observable or measurable differences in individual farm animals for a particular trait. This is the raw material with which the animal breeders work. If there were no variations among individuals, there would be no need to select or reject animals for breeding purposes, because they would all look similar and also perform alike. Phenotypic variation observed in animals in a herd can seldom be due completely to differences in genes. Such variation can be ascribed to heredity, environment, or the interaction of both. The importance and influence of each of these factors has been a very active topic of research and study in farm animals during the last several decades. An individual or a group or a family or a breed or a type of a species makes a special combination of special genes or gene groups at a particular point of time. It is, therefore, imperative to conserve all sorts of indigenous type of animal germplasm resources needed to be conserved, when needed again for improvement work. For example, in the Indian subcontinent, as many as 26 distinctly defined breeds constitute the distinguished zebu cattle germplasm resources. Each of these breeds has been evolved and successfully developed as per the special dictates and needs of their particular habitats.

India is considered as the original home of red jungle fowl. White Leg Horn (WLH) happens to be the most important and most effective egg laying breed in the world. Its main strains are replacing all other types of native poultry all over the world. Same is also true in respect of popularization of only two broiler types of strains. As a result of this type of dramatic development, poultry production both for egg and meat has been revolutionized in many parts of the world, and the indigenous breeds are still very active, hardy and can easily withstand adverse climatic conditions.

Our domestic farm animals and poultry have been serving mankind in multifarious ways since ages. There is a need to appreciate such valuable types of genetic resources available in many distinct types, species, breeds and forms. Some of them already face the danger of extinction and as such need to be conserved before it is too late. There is a great genetic diversity among the different breeds of animals in India. This diversity is a part of our national heritage, acquired during the process of domestication of animals resulting in their subsequent development into different types or groups, breeds, forms and strains.

Phenotypic plasticity

Increasing awareness of the nature of species as evident from population studies, has led to more meaningful investigations regarding their structural and biological characteristics. Both genetic and non-genetic phenotypic variation commonly occurs in the biology of sexual insects and many seemingly identical phenotypes may also conceal considerable variation at the level of the gene (Mayr, 2001). Gene frequencies differ among populations and environmental changes often result with varied expression of diverse genotypes. Therefore variation, whether genetic or non-genetic, form indispensable prerequisites of evolution, calling for a better appreciation of diverse aspects such as size variation, colour pattern, adaptive diversity and even molecular patterns as evident in the population of sexual insects. A knowledge of the patterns of such plasticities and an assessment of their evolutionary implications are useful in our understanding of species. Richards (1961) defines polymorphism as "one or more sexes of a species occurring in two or more forms which are sufficiently sharply distinct to be recognizable without morphometric analysis". The occurrence is regular or recurrent, the rare of the two forms makes up a reasonable proportion, the rarest type being essential for the survival of the species". Hence environments experienced by a population determine the phenotypes and establish their fitness for a particular trait. Phenotypic plasticity has to be regarded as a fundamental character of insects, subject to evolutionary pressure, not merely an intrinsic variation – an aspect deserving increased attention by those involved in studies on biological diversity wherein phenotyphic plasticity is an integral component. As such the ability of an organism to express different phenotypes depending on the biotic or abiotic environement, often referred to as *phenotypic plasticity*, plays an important role in the adaptation of insects, more especially in relation to their ability to thrive in changing environments. To be more specific, phenotypic plasticity is the expression of multiple phenotypes by one genotype allowing individuals the flexibility to alter their morphology, physiology and behaviour. In other words, polymorphism contributes considerably to our understanding of the casual factors underlying variation. Besides morphological specialization, is the broad tolerance of individuals in a variety of ecological conditions (*broad phenotypic plasticity*) as against that for a narrow niche. The existence of genetic polymorphism within species, producing individuals to a variety of ecologically adapted sub-niches is well known, besides the occurrence of ecotypic variation among populations within the geographic areas of area of species (Thompson, 1994). In short, groups of individuals within species have the potential to become specialized to exploit different habitats. For example, the realization that increase in humidity resulting with increase in altitude, predominantly influences pigmentation, so well illustrated by the lady bird beetle *Coccinella septumpunctata* Linn. (Kapur, 1959) (Fig. 5.3); or the common earwig *Labidura riparia* blackish brown in moist areas, but yellowish brown in desert areas, or life forms as in some Collembola involving progressive reduction in pigmentation, development of eyes and body size with increasing depth, to mention a few, has confirmed that each species is endowed with different structural and biological characters are revealed by population studies (Fig. 5.4).

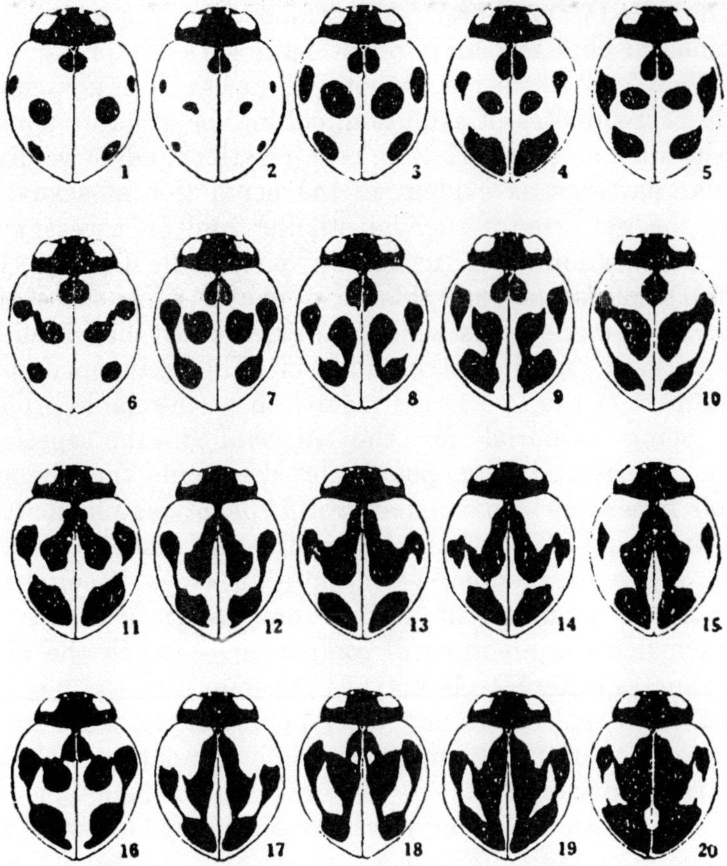

Figure 5.3. Pigment variation in a coccinellid beetle

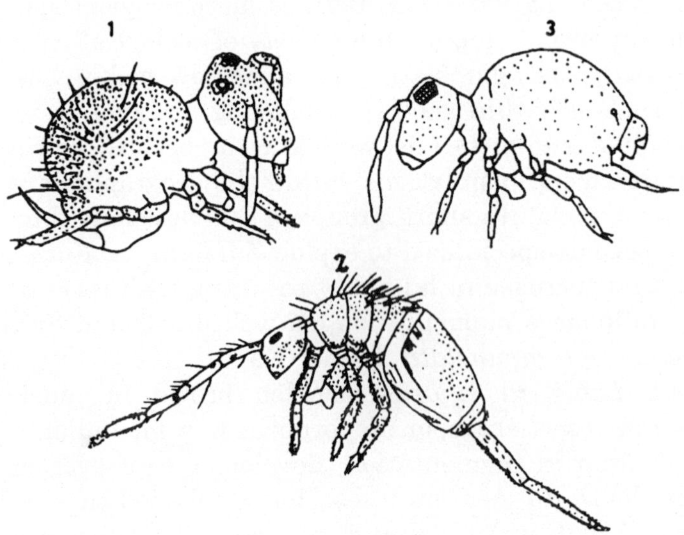

Figure 5.4. Collembola showing eye reduction

Figure 5.5. Solitary and gregarious individuals of *Schistocerca*

The degree of modification of the phenotypes varies with different groups. The phenotype in short, is a biological system resulting from the interactions of genotypes with the environment. The genotype does not give rise to the phenotype, but to a range of phenotypes. One of the finest examples of phenotypic plasticity relates to the locust *Schistocerca gregaria* which shows a solitary phase and a gregarious phase. In the *solitary phase* the nymphs are green, grey or brown and the adults show longer and crested pronotum and longer hind femur. The individuals are more dispersed and lead a solitary sedentary life. They breed in desert areas or outbreak areas during the recession period. This leads subsequently to the gregaria phase. In the *gregaria phase* the nymphs are black, yellow or orange, and the adults show shorter saddle-shaped pronotum and shorter hind femur. The individuals are active and crowded and they form swarms and migrate over great distances. The gregarious individuals are adults in structure physiology and behaviour (Fig. 5.7). Instances like increase in size as related to increase in temperature or the production of two or more discrete phenotypes in response to environmental signals as in some butterflies, have considerable implications. A change in phenotype occurring in response to such environmental signals necessarily results in changes in growth, survival or reproduction. An instance on hand relates to the impact of temperature on *Drosophila* resulting in a host of morphotypes (Fig. 5.7). Several such adaptive responses are known, such as the development of differently shaped mandibles in grasshoppers feeding on different diets, as against those appropriate to a given diet providing a distinct reproductive advantage. Since each species has varying biological characteristics, an analysis and comparison of these differences are essential prerequisites for identifying phenotypic sources of variation among individuals of a population. A good example is the genetically identical female morphs as seen in some aphids showing differences in host plant preference, morphology, size, number and timing of reproduction of offspring. The consequences of phenotypic plasticity therefore are diverse and an understanding of the functional impact of plasticity plays an important role leading to ecological and evolutionary success of organisms. Besides morphological aspects, such phenotypic variation can also occur in life history traits, size of egg clutches, larval growth rates, reproduction and survival.

Variant colour patterns in the Asiatic beetle *Harmonia oxyridis* are so striking that many of them were described as new species. The variation is caused by five alleles of a single gene; of these one gives the yellow and yellow black spotted form; another black yellow spotted; a third the yellow black remained and other two black, orange spotted. Similar variations are frequently seen in lady bird on coccinellid beetles which show pale to dark colouration with intermediate spotted patterns (Fig. 5.6).

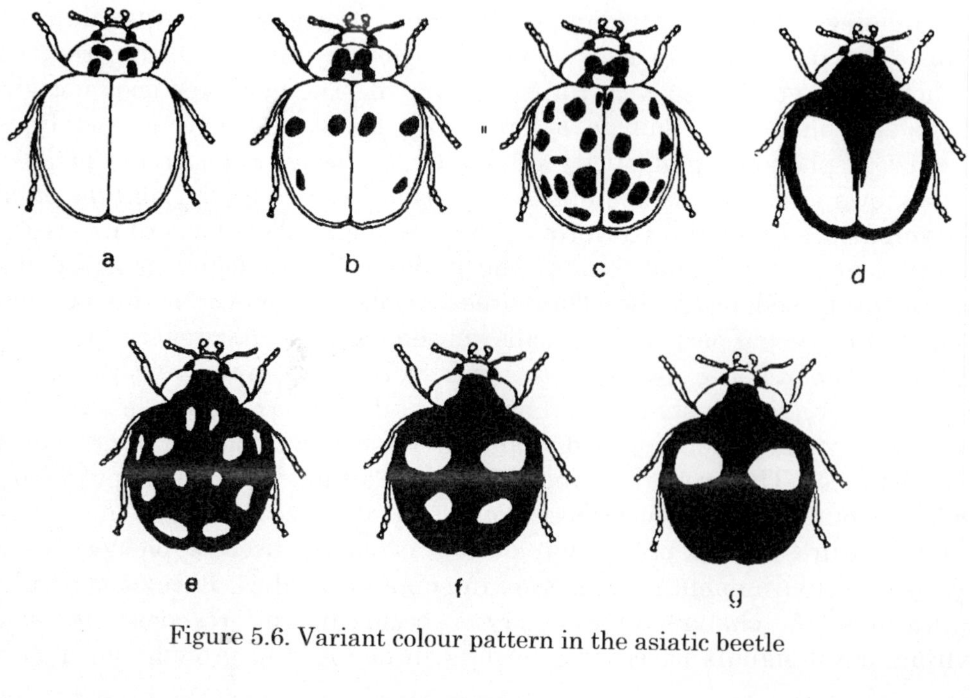

Figure 5.6. Variant colour pattern in the asiatic beetle

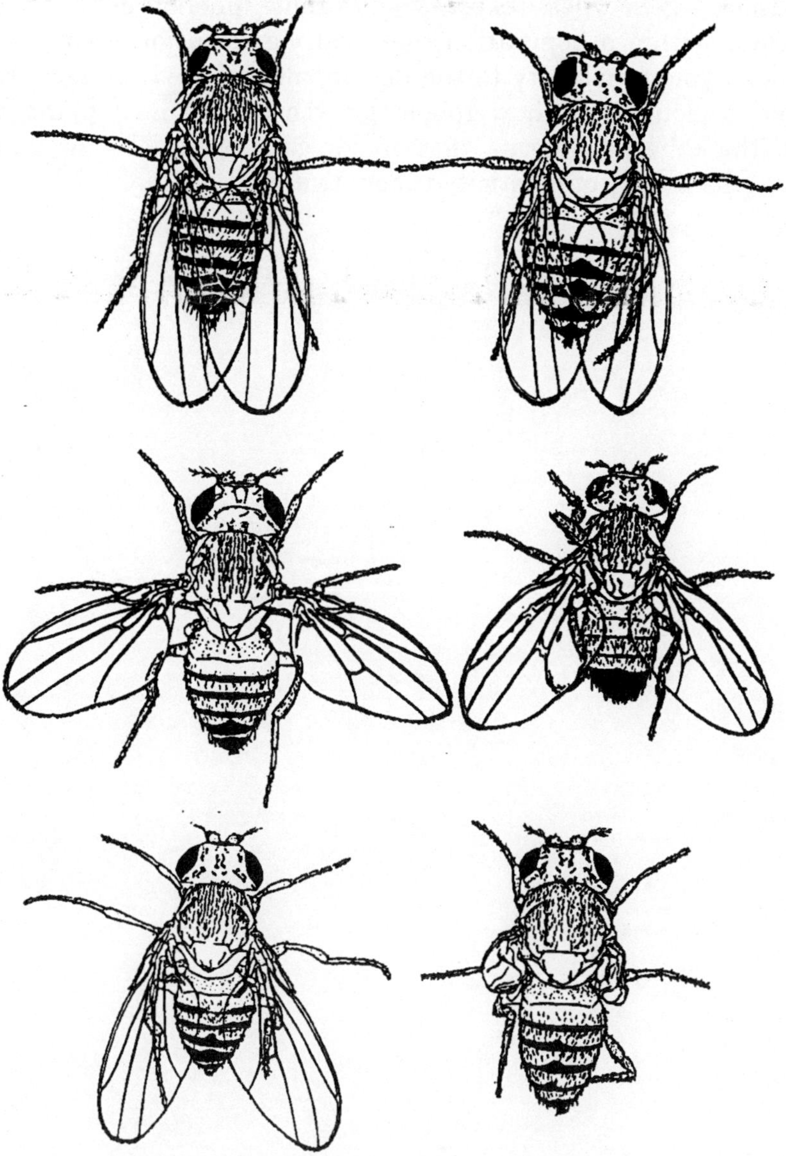

Figure 5.7. Induced variation in *Drosophila*

Similarly trophic polymorphism increasingly occurs in many fishes resulting from the effects of diet or early in development as in the development of snail-crushing morphs in some sunfishes, involving the development of different foraging tactics. The development of trophic morphs varies from phenotypic plasticity to genetic based polymorphism. Extreme effects of polymorphism occur in social insects like ants and termites with individuals specialized for foraging, others for defense and yet other to carry out intracolony duties.

Sometimes genotypes express different phenotypes in different environments. Within a population there can be variation among genotypes in their phenotypic sensitivity to the developmental environment. In aphids for example, a clone may show robust growth on one host plant, but poor growth on the other. In forms that make galls, plasticity in gall size is common and the degree of plasticity is genetically variable.

CHAPTER - 6

NUTRIENT CYCLES AND SOIL / LITTER BIODIVERSITY

The tropical forests have easily the highest production of organic matter with cyclic nature of nutrient inflows, the role of soil fauna and flora in the cycle being dominant. Long term stability of the forest ecosystem depends on the balance between the input and output of nutrients at various stages of growth of the forest. Nutrient cycles are therefore important, involving uptake by plants, their storage within the organism and subsequent return to the soil. Besides a rich variety of biological interactions exist in soil environments of natural and agricultural systems. Many of these interactions involve nutrient transformations and while some interactions have direct effects such as that of invertebrates and plants in soils, most other except their influence on plants indirectly through processes mediated by micro-organisms (Coleman *et al.*, 1983). Soil insects and arthropods play an important role in the break down of decaying plant material into nutrients available for uptake by plants. An understanding of the roles of organisms in energy flow and nutrient cycles is therefore of considerable importance.

Nutrient Cycles

The cyclic movement of the chemical elements of the biosphere between the organism and the environment are referred as *biogeochemical cycles* or *nutrient cycles*. Organisms utilize several chemical elements present in the biosphere in synthesizing new compounds which are essential for life activities. Although the biogeochemical or the nutrient cycle is superimposed upon the energy flow of the ecosystem, the two systems differ fundamentally in that, while the energy is derived from the single source, the sun, the nutrients are obtained from all the different regions of the biosphere like soil, ocean, or atmosphere. The contact zone between the lower atmosphere and the uppermost parts of land and ocean which harbours a good amount of organic life is the region of occurrence of the biogeochemical cycles. Biogeochemical cycles differ from the energy cycles in their direction of flow, being cyclic in the former and unidirectional in the latter. Energy gets degraded as it passes from one trophic level to another, while the concentrations of nutrients increase and once again reach the producers. The *decomposers* comprising mainly of bacteria and fungi form a major link in the transfer of nutrients back to nature. These *heterotrophs* release certain enzymes, as the dead and decaying matter and various basic nutrients of the cytoplasm which are products of this decomposition recirculate in the environment. The producers are fed upon by the herbivores and in the process they hand over not only the stored energy but also certain basic nutrients Herbivores when being fed upon by carnivores transfer the nutrients to the latter. Ultimately it is left to the decomposers again to free the nutrients by acting upon the dead and decaying bodies of the participants of the food chain. Small amounts of chemical elements are lost in the cycling and recycling processes. Nevertheless the biogeochemical cycles are a stable and steady phenomena. This is mainly due to their capacity for counterbalancing the loss by obtaining elements from other sources of the atmosphere. Basically there exist two types of biogeochemical cycles, the *gaseous cycles and sedimentary cycles*. The gaseous nutrient cycles are the nitrogen cycle and the carbon cycle although much of the total weight of organic matter of life is constituted by water, hydrogen, oxygen and carbon dominate, constituting about 90 per cent of the dry weight (Ananthakrishnan and Viswanathan, 1976; Ananthakrishnan, 1989).

In the biosphere there is a constant movement of elements. Living organisms obtain most elements as gases or as water-soluble salts. There exists a selective acceptance of these elements–those with high or low ionic potentials, being easily accepted rather than those with intermediate ionic charges. Of the chemical elements taken up by the organisms to form complexes of substances within the living systems, carbon and hydrogen are the most important. These enter into the composition of fats and carbohydrates besides helping in protein synthesis along with nitrogen. Sulphur is also a critical element often needed for the formation of various amino acids. Energy transfer in the cells is facilitated by phosphorus which also forms a major component of nucleic acids. Many other elements act as catalysts in the diverse chemical exchanges within the cells. There is a tendency for heat to be distributed more evenly through air, water and soil. As a result of changes in temperature of air, water and soil changes occur in the movements of ocean currents as well as in the circulation of winds. It is these that contribute to what is known as *hydrological cycles* wherein water is converted by heat into water vapour which is distributed by winds all over at random. In some places the moisture becomes precipitated by storms and the resulting water gets back to the ocean.

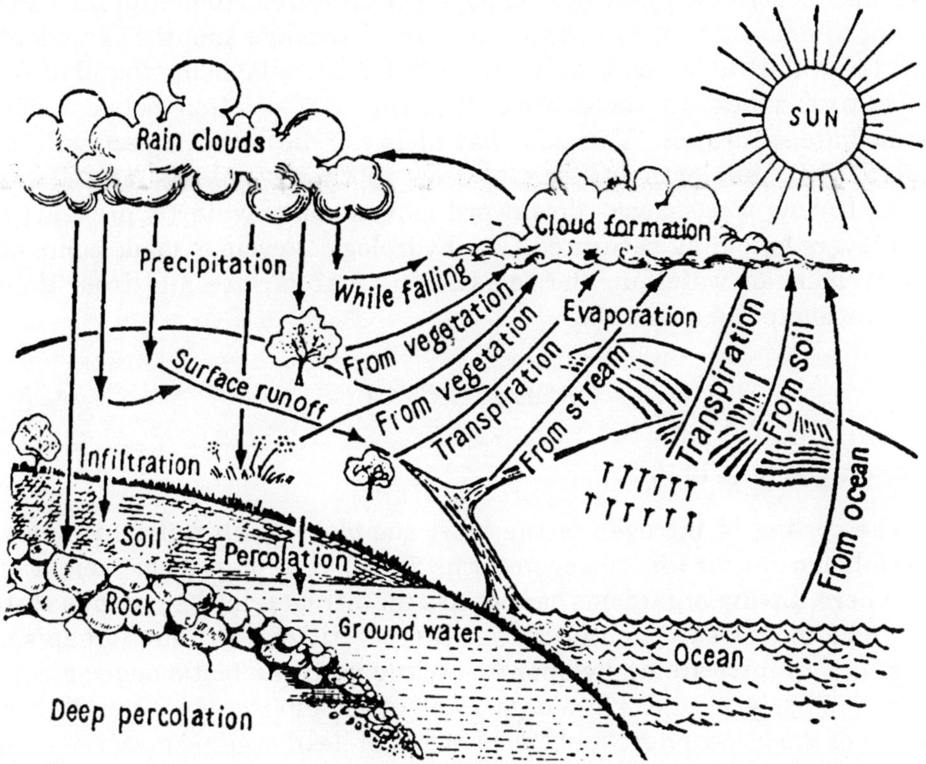

Figure 6.1. Hydrological cycle

Living organisms, atmosphere and earth maintain between them a circulation of water and moisture. This is designated as *hydrological cycle* (Fig. 6.1). The atmosphere possesses a limited capacity for holding water vapour, some amount of which is carried by warm air. The amount of moisture present in the ecosystem is derived from various sources. Mist, fog, ice, snow, etc. are produced by precipitation. Convection currents initiated by heating up of land and cyclonic disturbances are good examples. Plants derive water from the ground and some water also passes in from the atmosphere. But vegetation growing in semi-arid zones and deserts get their water from dew which results from the sudden lowering of temperature during night. Sometimes the precipitated water instead of going to the soil is returned to the atmosphere. This is because of *interception* by the plant communities. The degree to which it is effective is dependent on the nature of the existing vegetation and thermal conditions. Usually the precipitated water on reaching the ground goes to the soil and excess water runs off or gets evaporated. Soils with humus and litter exhibit a greater run off.

Evaporation from soil is accelerated by wind and high temperature. Force of gravity moves the penetrated water to the water table below. Hygroscopic forces tap the water to be absorbed by soil particles and to this, capillary forces also contribute greatly. Evaporation of water from capillary sources helps in the moistening of surface soil during summer months. The depth at which the water table lies is different in different soils, being found at deeper soils in arid zones. In these areas it occupies a position below a zone of intermediate saturation. This zone has plenty of dissolved oxygen with a free circulation of nutrients causing a deep rooting of plants and luxuriant growth. During wet seasons downward movement of water is prevented due to soil layers becoming saturated. The hydrological cycle is made complete by the movement of water up the roots to the leaf surface and from there by transpiration to the atmosphere.

GASEOUS CYCLES

Nitrogen Cycle (Fig. 6.2)

The cycling of nitrogen is the most complete of all nutrient cycles. An essential element for life, nitrogen forms 78 per cent of gaseous weight of the atmosphere. Living organisms cannot use up nitrogen in the gaseous state, in order to be converted into a form that can be utilized by living organisms, it enters into an intricate and elaborate cycling with the biotic environment. In the process bacteria and micro-organisms of the environment mineralize the nitrogen of the atmosphere and the nitrogen of dead organic materials.

Nitrogen is found in nature in the form of organic compounds like urea, protein and nucleic acids or as inorganic compounds like ammonia, nitrites and nitrates. The first step in the nitrogen cycle is the transfer of nitrogen from the atmosphere to the soil. Besides rain water which brings in a substantial amount of the element, nitrogen is added to the soil by the process of *nitrogen fixation*. The latter process is a more efficient one in its addition of nitrogen. Nitrogen fixation can be effected physico-chemically or biologically. Physico-chemical fixation is not significant. Biologically nitrogen fixation is established by the symbiotic organisms like bacteria and algae. The symbiotic fixation is well illustrated by the bacterium *Rhizobium*, which promotes nitrogen fixation only when associated with leguminous plants. Different strains of *Rhizobium* occur in different legumes and adherence to specific legumes is not uncommon. If separated from their host plants many bacteria show a loss of ability for nitrogen fixation. A leguminous plant attacked by *Rhizobium* shows development of nodules or outgrowths on the roots. These are formed by the penetration of root hairs by the free moving bacterial cells which increase to enormous numbers within the roots. The root nodules which are sites of nitrogen fixation often show a red or pink colour which results from the production of haemoglobin at these places. This phenomenon of nitrogen fixation is not entirely restricted to the Leguminosae. Many higher plants like *Casuarina* sp. with symbiotic bacteria, have been shown to have sites of fixation of nitrogen.

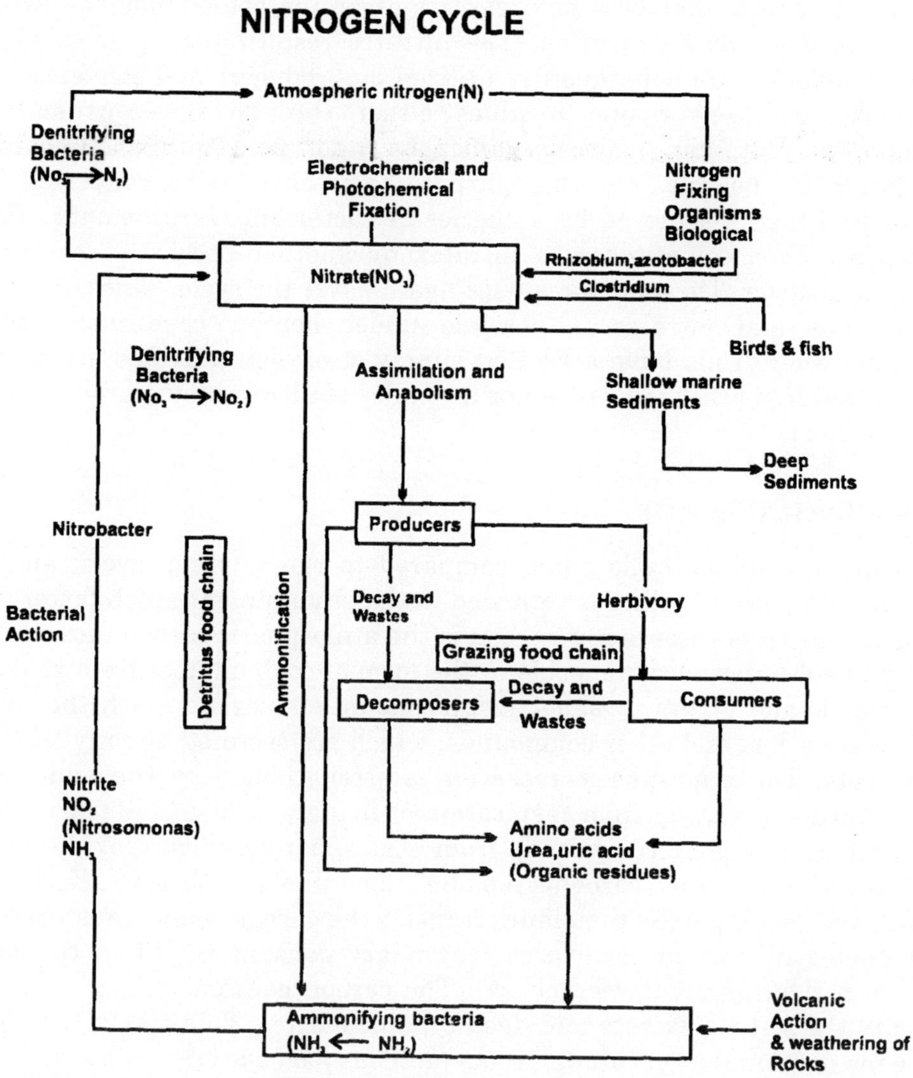

Figure 6.2. Nitrogen cycle

Predominant among the free living bacteria capable of nitrogen fixation are aerobic *Azotobacter* species or the anaerobic *Clostridium*. These are found in well aerated soils. The compounds of nitrogen resulting from the process of fixation are added to the pool of soil nitrogen. Organic residues such as amino compounds also result from volcanic action and weathering of rocks. Besides this, certain blue-green, algae found in aquatic environments also help in fixing up the nitrogen of the atmosphere. In all processes the inorganic nitrogen is fixed into an organic form and let out as waste products of metabolism which may be added directly or by the breakdown of nodules. Heterotrophic bacteria like *Bacillus* sp. and certain fungi utilize this nitrogen and convert it into ammonia or organic ammonium compounds the process being termed as *ammonification*. The nitrates resulting from physical and biological fixation are subsequently utilized by producers and converted into urea, uric acid, or organic residues either through the consumers or decomposers. But many micro-organisms cannot get at this ammonia or ammonium compounds. For this purpose they have to be converted into nitrites and this is achieved by a species of bacterium *Nitrosomonas*. These nitrites are then converted into nitrates by another species of bacterium called *Nitrobacter*. These nitrates once again enter the cycle. This conversion into nitrites and nitrates is possible under aerobic conditions and in situations where soils have a limited supply of oxygen, nitrates are quickly transformed into gaseous nitrogen or as oxides of nitrogen by a process called *denitrification*.

Carbon Cycle (Fig. 6.3)

This is a simple cycle when compared to the nitrogen cycle, and the natural carbon after cycling is returned to the environment much faster than nitrogen. The chief reservoir of carbon is the atmospheric carbon dioxide. The earth has subsidiary organic pools in the form of coal, oil, fossilized remains, and animals and plants. Besides, small amounts of carbon reach the soil as sugar, amino acid and other compounds, which are secreted directly from the plant roots. The atmospheric reservoir is strengthened by the addition of carbon dioxide resulting from respiration of living organisms. Many bacteria can synthesize organic compounds from atmospheric carbon dioxide. Plants (producers) take up carbon dioxide for photosynthesis. Herbivores (consumers) feeding upon the plants transfer the carbon compounds resulting from photosynthesis to carnivores (secondary consumers) when the latter prey upon them (*grazing food chain*). The carbon goes on cycling until it is added to the soil via excreta and dead organic matter. This enters the plants by means of assimilation through roots (*detritus food chain*).

The carbon cycle, although apparently simple, is complicated by the increase or decrease in concentration of carbon during the exchange between atmosphere and earth. The cycle nevertheless maintains a perfect equilibrium. The carbon present in the atmosphere and in the earth are compensated by each other when loss occurs at one site or the other. Lately carbon of industrial waste gases and gas products of household fuels have greatly enhanced the amount of carbon in the atmosphere to such an extent that a more or less imbalanced situation has resulted.

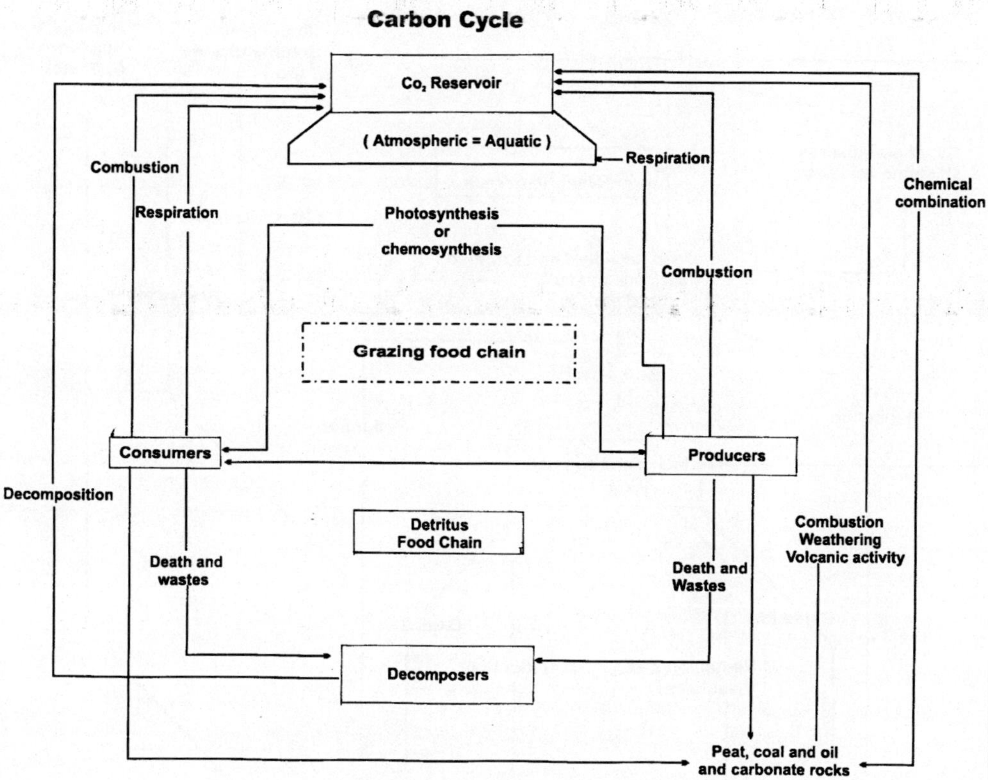

Figure 6.3. Carbon cycle

Sedimentary Nutrient Cycles

Most important among the sedimentary biogeochemical cycles are the sulphur cycle and the phosphorus cycle. These two elements are biologically important especially for the growth of living organisms.

Sulphur is present in the form of inorganic sulphates. It is reduced by bacteria into sulphides and sometimes it is present as sulphur dioxide or hydrogen sulphide. The latter is often lethal to normal life especially in aquatic environments, where its production is more due to the decomposition of detritus. Sulphate present in water is reduced by autotrophic plants and is incorporated in the amino acids. Hydrogen sulphide is released as the bodies and plants undergo decomposition and also through the excreted faecal matter of grazing animals. Certain specialized bacteria act on hydrogen sulphide to convert it back to sulphate. But what usually happens is that this hydrogen sulphide instead of being oxidized is passed into the media as such due to anaerobic conditions that may develop, since it is in sedimentary regions that this conversion takes place. Iron present in the sediments combines with sulphur to form the alkaline insoluble ferrous sulphide. But in aerobic conditions it releases sulphur (Fig. 6.4).

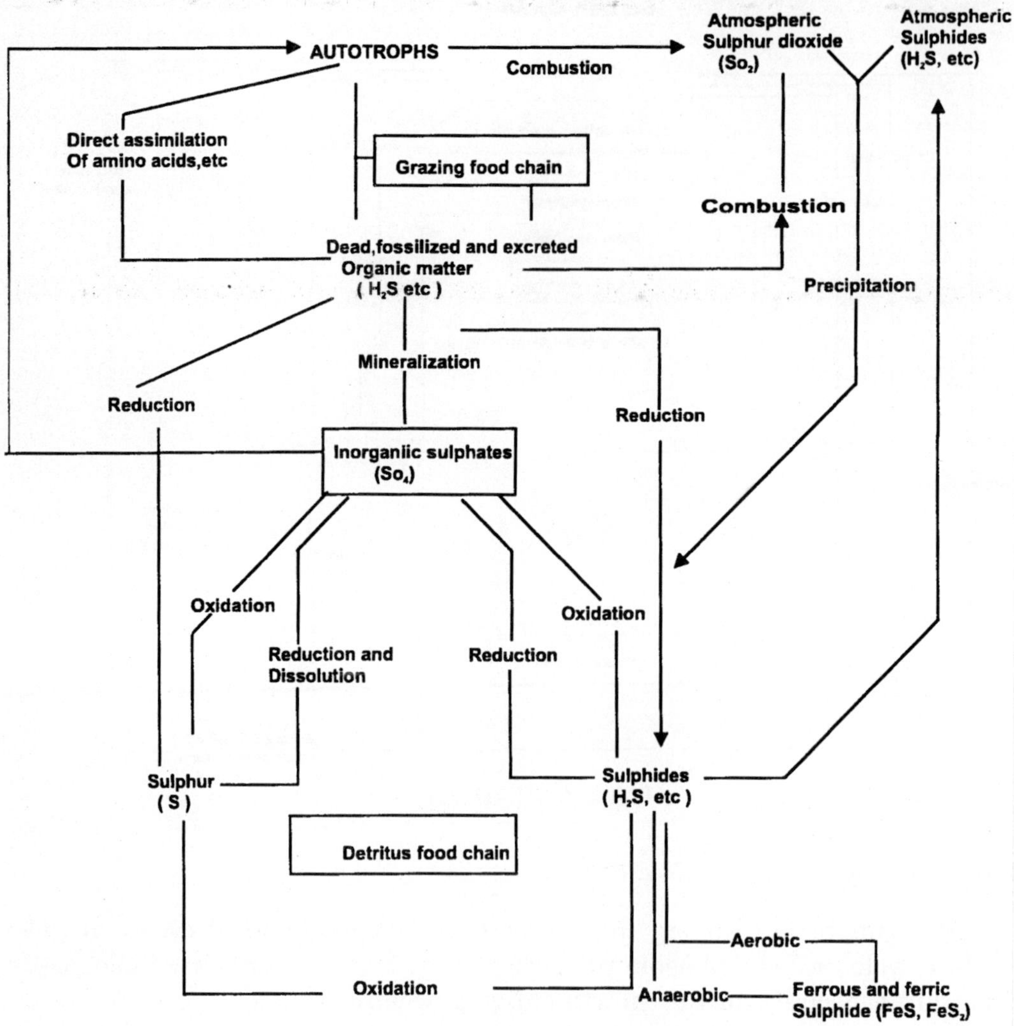

Figure 6.4. Sulphur cycle

Phosphorus Cycle (Fig. 6.5)

Phosphorus may be present in the environment in the form of organic mineral salts or as organic chemical compounds. Plant roots absorb phosphorus in the above mentioned conditions from the soil. These are then passed to the plant body and eventually to herbivores which feed on them. When these organisms die the phosphorus returns to the soil due to the decomposition of dead matter. Under many circumstances inorganic intake is supplemented by the release of a large amount of organic phosphorus by the breakdown of excreta and dead material. The amount of phosphorus circulated is highly variable in different seasons, but summer induces a larger circulation. Fertilizers added to the soil combine with the soil phosphorus and cause depletion in the amount available to plants. But the micro-organisms of the soil act on the sugar exudates of the roots converting them into organic acids. These organic acids enter into the transformation of phosphorus into orthophosphates and many specialized micro-organisms release the inorganic phosphate from the organic phosphate compounds.

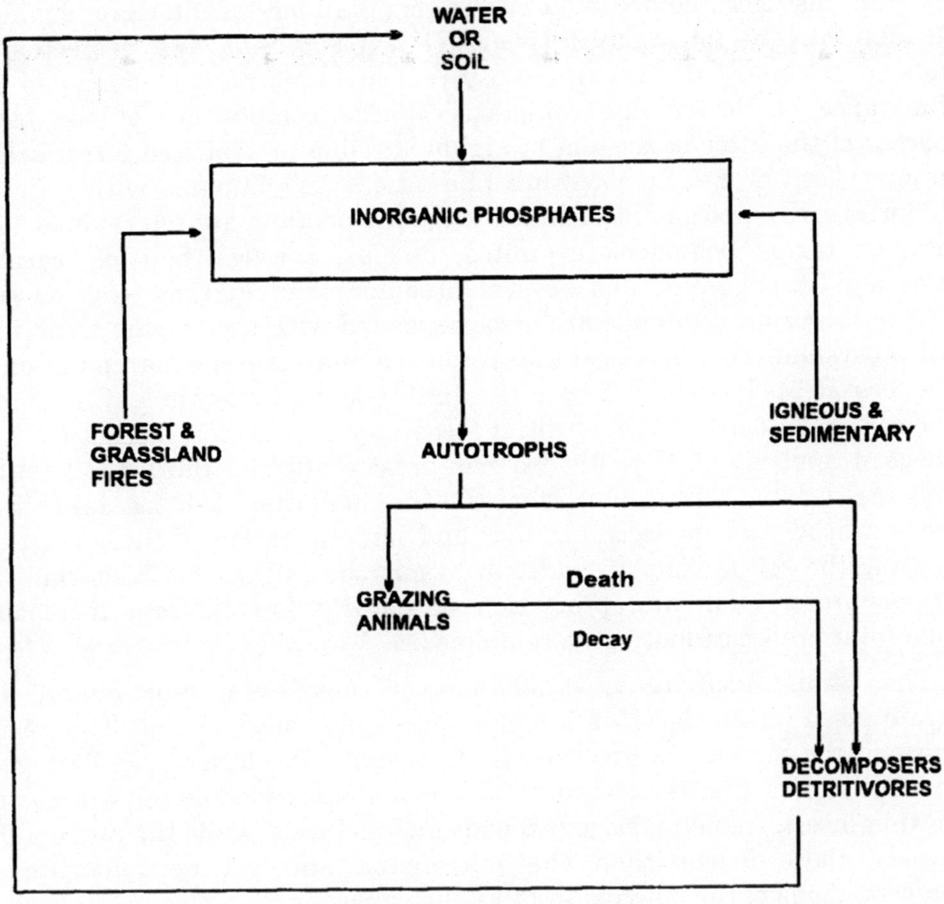

Figure 6.5. Phosphorus cycle

Sea water tends to become progressively filled with more phosphates brought by streams carrying phosphorus released through erosion. Wind transportation of salt spray carries some phosphorus from the sea back to streams. Birds and fishes excrete excess phosphate in their faeces and the huge accumulation of faecal deposits from guano gives back phosphorus to the continents. Terrestrial ecosystems more often lose phosphorus due to the absorption of suspended phosphorus by oceanic sediments.

SOIL AND LITTER BIODIVERSITY

The usefulness of forest litter ecosystem which is enriched in potential natural resources and their role in biodegradation is well known in recent years. Sufficient work has been initiated to identify these potential resources which are of vital importance in understating the nature of biological systems. It is a well known fact that natural forest ecosystems support a variety of consumer and decomposer systems than forests interfered by man, both qualitatively and quantitatively. The preservation and protection of forests to safeguard the natural ecosystem are important elements in the maintenance of the trophic complexities. The composition of the biotic resources of the litter ecosystem is highly variable in exploited forest areas, man-made forests and monoculture plantations as compares with uninterfered forests. Examples of litter inhabiting entomofauna are referable to such insects as thrips, collembolans, mites, psocids, earwigs, blattids, carabid beetles and staphylinids, and several predatory hemipterans such as geocorids, anthocorids, reduviids are also associated with the trophic chain in a forest ecosystem. The increased interference of man and his manipulation of forest ecosystems has reduced both the quality and quantity of leaf litter and in turn the entomosources of litter ecosystems. Further, increased and haphazard application of pesticides has greatly affected the litter habitat, largely disturbing the trophic chains. The majority of litter inhabiting microarthropods aid in decomposition and mineralization of litter, thereby increasing the soil fertility in addition to microbes, fungi and bacteria. The need for preservation and protection of natural forests from increasing human interference cannot be overemphasized.

The terms *microfauna, mesofauna* or *meiofauna, macrofauna* are generally used in the classification of soil animals based on size. The micro, macro and the mesofauna are typical of the soils, the former hardly bigger than a millimeter, live in micropores. The macrofauna include the worms and borrowing insects, nematodes, arthropods and molluscs, while the microfauna comprises those bigger than the microfauna and yet not elongate or vermiform. Some of the macrofauna like the earthworms, play a useful role in the decomposition of organic matter and hence in humus formation. Arthropods like insects, millipedes, centipedes, spiders, larvae of flies, ants and termites help in soil aeration through burrowing, feed on decaying vegetation and in some cases play a useful role as predators in the soil food chain. Slugs and snails are more abundant of soil molluscs and act as scavengers and are better equipped to digest cellulose.

The faunal elements living in water-filled pores are the *hydrobionts* and those living in air-filled pores are the *aerobionts*. Hydrobionts comprise most of the microfauna, protozoa and nematodes. Aerobionts live in bigger soil cavities and in view of changes occurring in proportion of air in soil, say after rains, there are adaptations wherein some of the smaller insects are effectively protected by capillary action and by the formation of strings of air bubbles. So even the aerobionts have to be moisture lovers or hygrophiles. The chemical composition of the soil is very vital for the fauna, such as calcium, which though of not much use to insects, is nevertheless important to molluscs, millipedes and earthworms.

Biotic resources of the soil

A healthy soil is vibrantly active or alive teeming with bacteria, fungi, moulds, yeasts, algae, protozoa, worms and insects living in the top cores. Soil bacteria occur in clumped colonies around individual soil particles forming 0.03 per cent of the weight of the top soil and the roots and root hairs of plants are generally enclosed in a bacterial film and they are more dense in the rhizosphere. About four billion bacteria are estimated per gram of soil, and in good soils bacterial matter has been known to weigh about 5600 lb/acre. More than 20 million actinomycetes (ray fungi or thread bacteria) occur in the soil, the estimated number being one million fungi in one gram of dry soil, their biomass per acre reaching 1,200 pounds. Algal forms such as diatoms, blue, blue-green, yellow, green, occur in high densities of dry soil. They aid in soil aeration through release of oxygen, besides serving as food for other soil organisms. The teeming variety of animal life in the soil is equally significant. Table 1. gives a rough estimate of the abundance of living organism in soil.

Table 1. Abundance of soil organisms

Groups	Abundance
Bacteria	1,000,000,000/gm
Actinomycetes	5,000,000/gm
Protozoa	500,000/gm
Algae	200,000/gm
Moulds	200,000/gm
Nematodes	175,000-20,000,000 sq m
Molluscs	50,000/acre
Myriapods	1,000,000/acre
Annelids	1,000,000/acre
Arthropods	1,000,000/acre

(After Southwick, 1976).

Soil inhabiting protozoans like amoeba, ciliates, zoomastigine flagellates occur near the surface soils, while the testate forms like *Thecamoeba*,

Euglypha, Difflugia, have a wider vertical distribution. *Bipalium* is a common example of a terrestrial polyclad, while among nematodes, species of *Rhabditis, Diplogaster, Tylenchus, Heterodera, Aphelenchoides, Mononchus* abound by as much as 1-3 million in raw humus soils to 20 million/m^2 in grassland soils. Among annelids the microscopic enchytraeids are represented by *Enchytraeus fridericia* and *Achaeta lumbricellus,* which are more abundant in organic soils and forests rather than in grasslands. Earthworms have been proverbial for their influence on the process of decomposition of organic material, breaking up litter fragments and mixing them thoroughly with mineral soils resulting in the formation of organic mineral soils. Some of the common Indian species are those of *Megascolex, Pheretima, Octochaetus, Drawida* and *Moniligaster*. In dry soils, earthworms are completely absent, more so the surface dwellers since this layer is subject to desiccation. Species like *Lumbricus terrestris* are deep burrowing and soil moisture as well as pH have a limiting effect on the distribution of earthworms. Surface dwellers generally undergo vertical migration in summer. The population size of earthworms also varies with the nature of the habitat, maximum densities occurring in alkaline grassland soils and sparse incidence in acidic soils. It has also been observed that surface dwellers produce more cocoons than deep burrowing forms. Earthworms feeding on plant litter and other plant debris in the organic matter rich soil surface are called the humus farmers, comprising the epigeic and anecic forms. While epigeics are surface dwellers and phytopathogous, anecics feed on the leaf litter mixed with the soil of the upper layers. Endogeic forms live deep within the soil deriving nutrition from the organically rich soil they ingest. Earthworms and the entire volume of soil litter they influence through their activities constitute the drilosphere (Fig. 6.6) (Ismail, 1997).

It is among the arthropods that we come across the maximum number of surface dwellers. These involve a limited number of crustaceans like the isopods and amphipods, but mostly the myriapods such as symphylids, millipedes and pauropods; among the insects there occur primarily the apterygote, proturans and collembolans among the pterygotes adults and larvae of Coleoptera and Diptera are common; arachnids comprise mostly of mite populations. One of the most successful groups of crustacea inhabiting the soil are the isopods and to some extent the amphipods. Isopods are more tolerant to desiccation than amphipods which mainly occur in decomposing plant materials. *Oniscus, Porcellio, Armadillidum* are the most common isopods of the tropics in the humid zone and are very susceptible to changes in variation of moisture since they have a permeable cuticle.

Symphylid myriapods are very important components of the soil fauna occurring in a variety of soils, mostly moist organic soils, and undergo seasonal vertical migrants pauropods with their characteristic trifid antennae rarely exceed a few hundred per sq. meter and are seen in damp leaf litters. Of the chilopods the carnivorous *Scolopendra* and *Lithobius* are common in moist soils feeding on leaf litter inhabitants. Millipedes or diplopods are better represented in the forest floor by species of *Spirostreptus, Thyropygus, Glomeris, Arthrosphaera, Polydesmus, Iulis,* etc. and their large size and litter feeding habits contribute considerably to the process of decomposition of organic matter, leading to the production of humus.

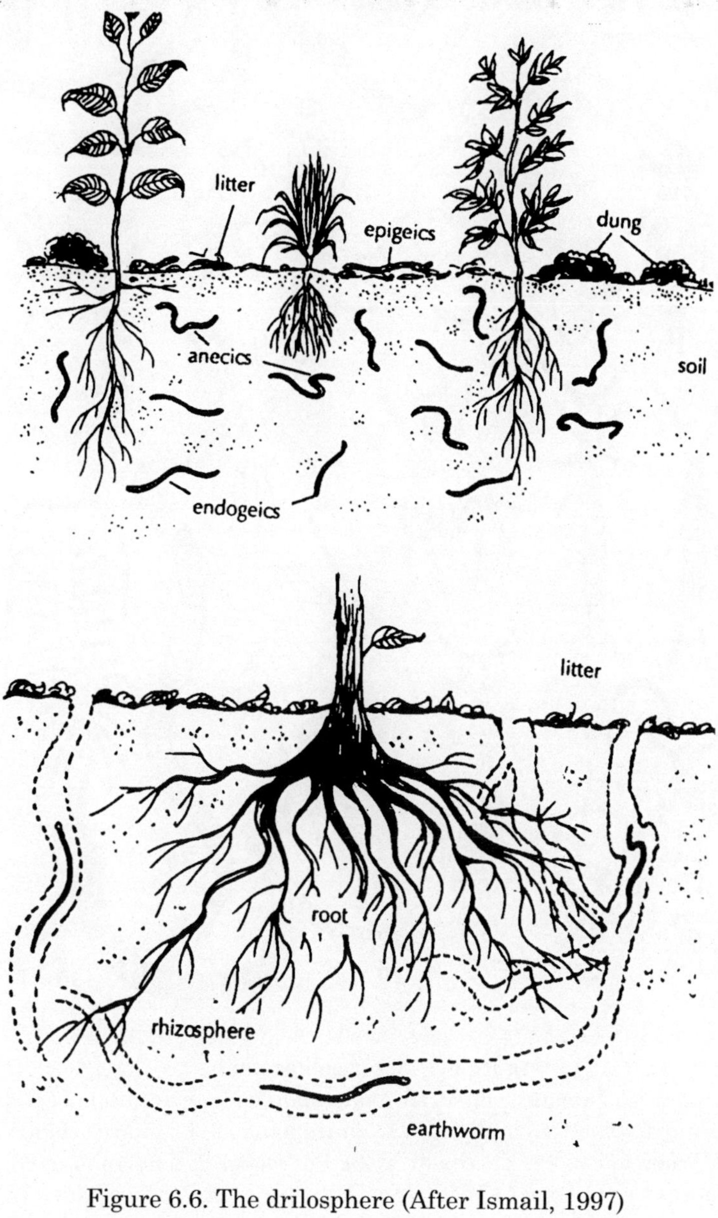

Figure 6.6. The drilosphere (After Ismail, 1997)

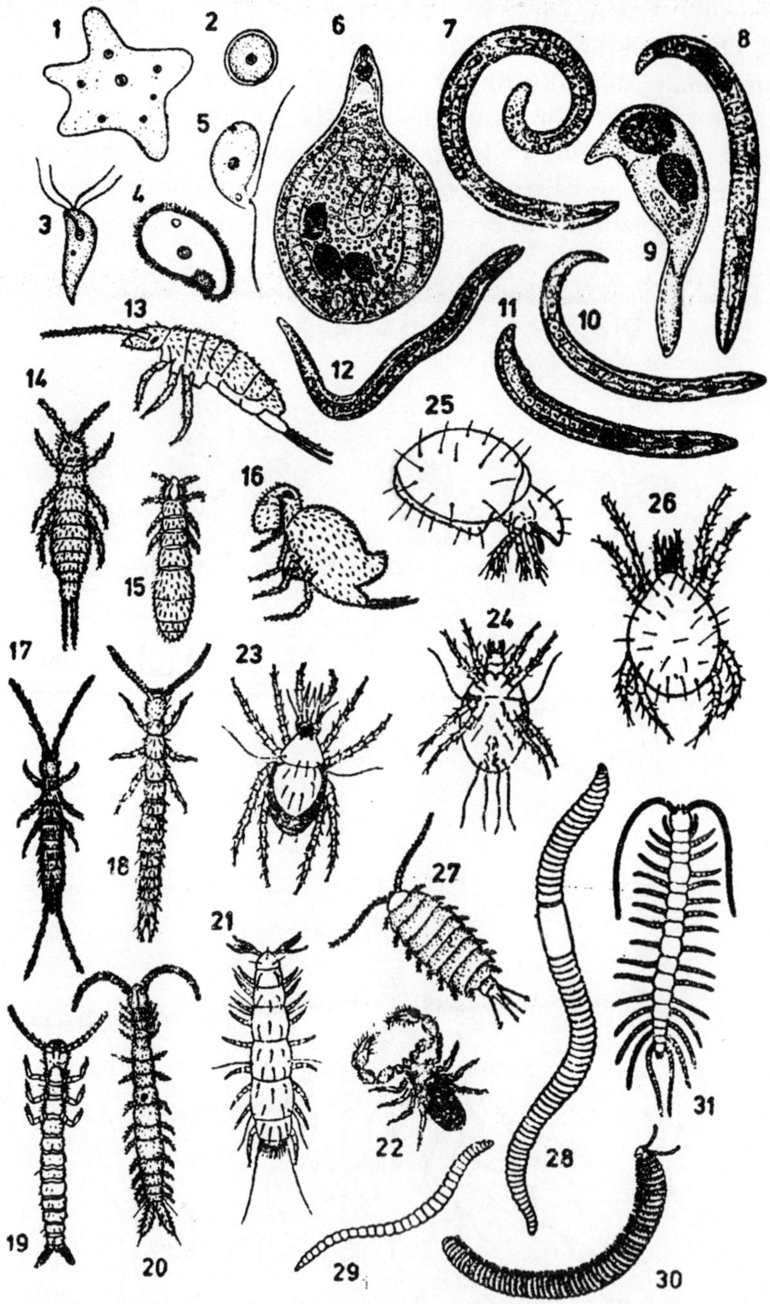

Figure 6.7. Some soil animals.
1-5. Protozoa; 6-12. Nematodes; 13-16. Collembolans (springtails); 17. Campodea; 18. Japyx; 19. Anajapyx; 20. Symphylid; 21. Pauropods; 22. Pseudoscorpion; 23-26. Mites (Acarina); 27. Wood louse; 28. Earthworm; 29. Pot-worm (Enchytraeus); 30. Millepede; 31. Centipede (*Courtesy* CSIR, New Delhi, *Science Reporter,* July 1973).

Insects abound in soil and litter and it would be next to impossible to indicate the relative abundance of each species of the several orders represented. Broadly stating, the apterygotes, Collembola and Protura form numerically the most important groups of soil insects. The Protura, Diplura and Collembola are the major soil dwellers. Japygids like *Anajapyx*, *Japyx* and *Campodea* are often found in small numbers in moist soils under stones and in humus and litter layers. Proturans are more abundant than Diplura and very common in moist forest and grassland soils which abound in species of *Eosentomon*, *Acerentomon* and *Acerentulus*. Of all the soil microarthropods, Collembola are the most predominant and stand comparison only with mites. The Onychuridae, Isotomidae, Poduridae and Entomobrydae are richly represented in the soil both in number and species composition. The type of soil, the composition of microflora, soil moisture and the type of vegetation influence the abundance of Collembola, although variation in pore size and organic content equally influence their variation in density in different soil types. Large forms like *Tomocerus*, *Entomobrya* and *Orchesella* abound in surface litter, while the smaller *Onychurus*, *Tullbergia*, etc. occur in deeper layers. In the top 10-15 cm, Collembola abound in grassland and forest soils, because this is the region where active breakdown of organic material takes place. The diameter of the pore cavities in this region is smaller and the moisture content of the organic layer is also less. *Life forms* have been characterized on the basis of distribution of Collembola involving progressive reduction in pigmentation, development of eyes and body size with increasing depth. (Fig. 6.7)

Others like Dermaptera, Psocids, Dictyoptera, Isoptera, Coleoptera, a few Hymenoptera and some Diptera also occur in the soil, sometimes as juveniles. The termites such as *Reticulotermes*, *Odontotermes* are important soil dwellers of tropics and play an important role in the break up of organic materials and their mixing up with mineral soils. Among the Coleoptera the carabids and staphylinids and to some extent cicindellids or tiger beetles are associated with soils principally along with a few others, like the larvae of click beetles or elaterids. Larvae of nematoceran Diptera like those of tipulids, mycetophilids, the brachyceran tabanids, asilids and empidids, and the Cyclomorpha represented by the muscids, phorids, the carrion feeding calliphorids and sarcophagids often abound in the soil. Among the Hymenoptera, ants are the most important soil dwelling forms though a few bees and mutillid wasps are also known to inhabit soils. Ants abound in forest litter and cultivated soils. Representatives of *Camponotus*, *Cremato-gastor*, *Formica* are wood dwelling and many members of Myrmiciniae build mounds. The availability of food, the physical nature of the substratum as well as moisture content of the soil are important factors limiting the distribution of ants.

Mites are the most abundant in organic soils and are represented by the cryptostigmatid detritus feeders; the mesostigmatid and prostigmatid mites include predators and detritus feeders. The Astigmata generally occur in pasture soils. Mites flourish in moist organic soils and forms like *Galumna*, *Cepheus*, *Hemorobates* have been recorded as inhabiting exposed situations such as in lichens, mosses, etc. The detritus feeding mites have been known to withstand varying conditions of environment, such as moisture organic content and pH. They are often rare in cultivated soils and generally prefer undistributed surface organic layers. The large forms like *Macrochaetes*, *Pergemasus* and *Steganocarus* are generally common in surface litter, while the medium sized and smaller forms like *Scheloribates* and *Brachychthonus* penetrate into humus and even deeper. Tradigrades or bear animalcules also occur in surface layers of moist soils in grasslands, being represented by the more common genera *Macrobiotus* and *Hypsibius* (Ananthakrishnan, 1996).

CHAPTER - 7

HOST-PARASITE-VECTOR BIODIVERSITY

One of the substantial facets of plant, microbial and animal diversity is the spectrum of parasites, whose occurrence, distribution and morphological, physiological and behavioural adaptations are closely linked with their respective hosts. Parasitism can best be defined as that relationship between two organisms, in which one organism lives on or in another organism called the host utilizing the latter for shelter and food. Most often the parasite harms the host; in other words it is pathogenic to the host. In the course of evolution, host-parasite relationship has become refined and fine tuned so that the parasite engages in subtle exploitation of the host by becoming non-pathogenic (not causing harm to the host) e.g. *Entamoeba gingivalis* in the gums of teeth in the buccal cavity of humans. Normally, every parasite has its own specific host and failure to discover the host leads to the death of the parasite. Several complications may be involved in the life histories of some parasites, which depend on particular intermediate hosts to complete their life cycles. For example, the malarial parasite spends part of its life-cycle in the stomach of a mosquito (*Anopheles*). The same is the case with the filarial worm, which also depends on another species of mosquito (*Culex*).

Several degrees of parasitism can be recognized. According to their position either outside or inside the body of the host, they are termed as *ectoparasites* and *endoparasites*. Ticks, fleas, mites and lice are examples of ectoparasites, while the round worm inside the alimentary tract is an endoparasite. The itch mite (*Sarcoptes scabei*) commonly burrows in the skin of man causing severe irritation. Parasites like the malarial parasite (*Plasmodium*) which can exist only within, a host are termed *permanent parasites*. Some individuals though free living as an adult, pass through a transitory parasitic stage as in the case of the sheep nasal fly. Its larvae are parasitic in the nasal cavity of sheep and cause serious disease. This is an instance of *temporary parasitism*. Parasites which are forced to lead the life of parasites are known as *obligatory parasites*. Others have the capacity to exist as a free living individual, but yet be parasitic when no food is available. These are the *facultative parasites*. While it is impossible to give a brief, yet accurate definition of parasites, in view of their endless adaptations, the following quotation nevertheless presents a good picture of a parasite: "A parasite is a living organism, which establishes a physiological association with the tissues on the surface of, or inside the body of another organism, which is usually bigger and stronger than the parasite and always belongs to a different species, the purpose of this association being primarily to provide food for the parasite; this other organism is called the host of the parasite and the host always reacts to some degree against the parasite and the injury it suffers". Host specificity varies enormously; while some parasites are associated with literally hundreds of different host species, others are restricted to only a single species.

The parasitic animals are confined to only a few phyla of the animal kingdom, namely, the Protozoa (e.g. *Plasmodium,* the malarial parasite, *Entamoeba,* the causative organism for amoebic dysentery), the Platyhelminthes or flat worms (tape worms and liver-flukes), the Nemathelminthes (different kinds of round worms), the Annelida (mostly external parasites like the leeches), and the Arthropods (including several insects like the mosquitoes, houseflies, bugs, lice and fleas). Some parasites like the round worms spend their entire life in a single host, while others like the malarial and filarial parasites depend on two different species to complete their life-cycle. In such cases, the terms *permanent host* and *intermediate host* are used. In the case of the malarial parasite, the mosquito is the primary or permanent host in which the parasite completes its sexual phase, and man is the intermediate or secondary host in whom, the asexual phase of the parasite is completed. Among Protozoa, *Trypanosoma gambiense* causes sleeping sickness and the parasites are transmitted by a fly, the Tsetse fly. Similarly Kala-azar or black fever is caused by a parasite *Leishmania donovani* transmitted by sand flies. The house flies transmit *Entamoeba histolytica,* which causes amoebic dysentery. *Giardia lambida* causes giardiasis. Similarly several species of coccidians are responsible for coccidiasis in poultry birds, Myxozoa for fish diseases and Microspora for pebrine disease in silkworms.

The bulk of the worms fall into two main categories, the flat worms and the round worms, or technically they are referred to as the *helminths*. The flat worms (Platyhelminthes) are more lowly organized, flattened from above downwards. The parasitic forms show a reduction or an absence of the organs of locomotion, sensory and nervous systems. The round worms on the other hand, show a higher grade of organization, the body being generally cylindrical and elongated, more or less pointed at either end. The two sexes are mostly separate and not bisexual as in the flat worms. Helminths are a source of great nuisance to mankind, because of their parasitic habits, almost always resulting in economic loss, sometimes even death. The parasitic flat worms belong to two categories, the flukes and the tape worms. The flukes are flat and leaf-like, while the tape worms are elongated and ribbon-like.

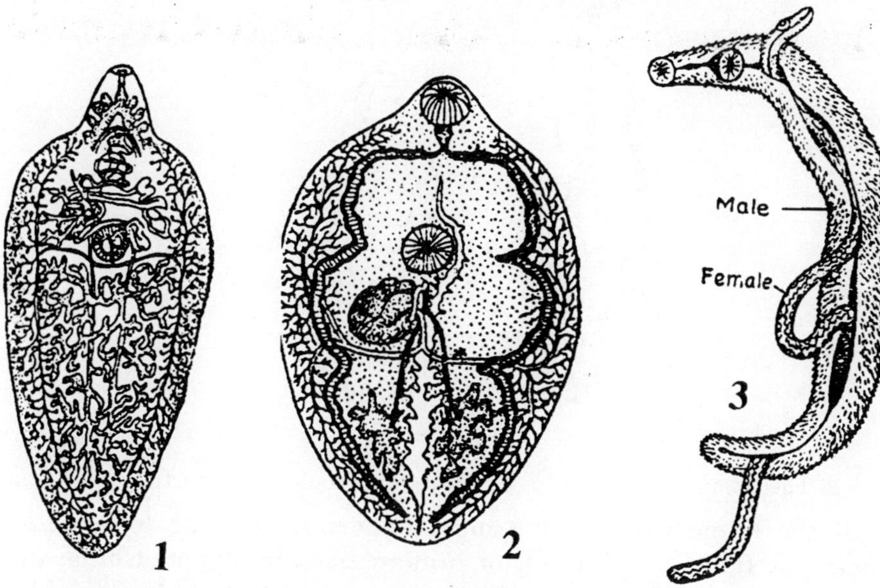

Figure 7.1. Some parasitic flukes. 1,2. Liver thinker; 3. Schistosomi

Flukes, Tapeworms and Roundworms

Flukes may inhabit the liver, the intestines, the lungs and blood. Accordingly, they are designated liver-flukes, intestinal flukes, lung-flukes and blood-flukes respectively. Intestinal flukes are exceedingly rare in man and those that are existent are accidental, in view of man's varying food habits. Suffice it to say that two of them are reported from India (Assam) *Gastrodiscoides* and *Fasciolopsis*. These inhabit the caecum and large intestine of man and cause inflammation and diarrhoea. More than a million cases have been reported from Assam. The well-known liver-fluke is *Fasciola hepatica* inhabits the bile ducts of sheep and other mammals. Human infections are known, but are comparatively rare. It is, however, very destructive to sheep, causing a disease known as the 'liver-rot' to which millions of sheep succumb. Lung flukes are very rare in man. However, *Paragonimus* has been found to affect man, chiefly through eating raw crabs or ray fish, which act as intermediate hosts. This is common in countries like Japan and India.

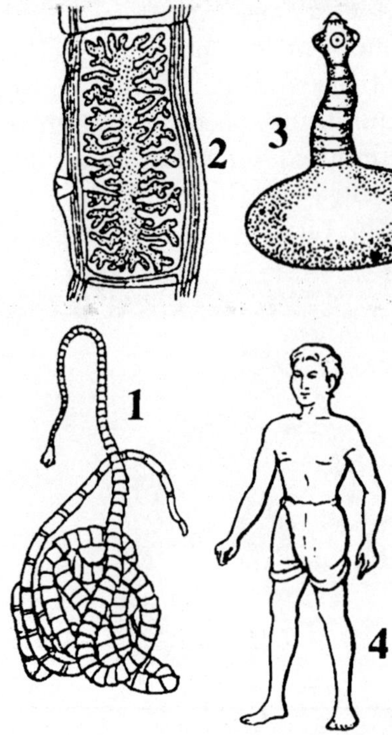

Figure 7.2. The human tape worm: 1. adult, 2. ripe proglottid, 3. cysticercus, 4. man

Of all the human flukes, the most important are the *blood-flukes* or *schistosomes.* It takes a great toll of human lives in Egypt, China, Japan, Africa and India and other oriental countries. *Schistosoma haematobium* is common in India, Africa and South West Asia. It lives mainly in the blood vessels of the urinary bladder and adjacent areas. It causes a serious disease called 'schistosomiasis' or '*bilharziasis*', characterized by blood in urine (*haematurea*). The rate of mortality is high, there being enormous enlargement of the abdomen and general emaciation. The sexes are separate. The female is more slender, longer than the male and is seen to lie in a longitudinal groove in the body of the male. This canal is called the *gynaecophoral canal.* The female is able to execute quick movements into the blood capillaries to deposit her eggs and returns back to the canal of the male. Oral and ventral suckers are present in both males and females (Fig. 7.1).

Tape worms are flat worms with long, flat, ribbon-like bodies; some of them like *Taenia solium* inhabit the intestine of man. The head of the worm is about the size of a pin's head and is the organ of attachment to the walls of the intestine. It is called the *scolex.* It bears a small prominence which is protrusible and called the *rostellum.* A double ring of chitinous hooks and four suckers are present at the base of the rostellum. An interesting feature is that in all tape worms, an alimentary canal and mouth and an anus are totally lacking. They absorb the predigested food in the intestine, where they live. Further, the bodies of all tape worms are made up of a number of units, the *proglottids*, each of which contains a complete bisexual set of reproductive organs. The posteriormost proglottids contain fertilized eggs and are termed ripe proglottids. Each proglottid carries laterally a genital opening. The common tape worm in the human intestine are the *pork tape worm*, common among pork eaters and the *beef tape worm*, among beef eaters. Digestive upsets, abdominal pain, nausea, weakness, etc. are common symptoms of tape worm infections. Mention may be made of a minute tape worm *(Taenia echinococcus)* 3-8 mm long, commonly inhabiting the intestine of dogs. Human infection is through kissing dogs, allowing them to lick faces and such close associations. In man, they develop in the liver and the larvae develop into cysts, which grow bigger and bigger and after a few years reach a considerably large size causing pressure on adjacent organs and the results are very disastrous (Fig. 7.2).

Round worms

The round worms far outnumber the flat worms in their abundance and it has been estimated that there are as many as 80,000 round worms, parasitic in vertebrates alone. Some of the important round worms parasitic in man and which are of vital importance, as causative organisms of disease and suffering, are listed below :—

(1) The hook worm (*Ancylostoma duodenale*).

(2) The common round worm (*Ascaris lumbricoides*).

(3) The Filarial worm (*Wuchereria bancrofti*).

(4) The Trichina worm (*Trichinella spiralis*).

(5) The Guinea worm (*Dracunculus medinensis*).

The *hook worm* is a very dangerous human parasite, causing him great economic loss, not to say of fatalities. The adult hook worm inhabits the small intestine, feeding on the blood and tissue juices. The number of eggs produced by the hook worm is very large, and they pass out along with the human faeces. They hatch into tiny larvae pointed at either end. In this stage, they enter the human host through the pores of the skin, when they walk through dirty water or when they come into contact with damp earth. Little children, habitually playing, in muddy puddles, often get infected.

The largest of human round worm is *Ascaris lumbricoides*, a very common intestinal parasite especially among children. The males reach a length of 6'-12' while the females are about 8'- 18' long. Though it is supposed to feed on the semi-digested food, it is also known to lacerate the mucous membrane and suck blood. The number of eggs produced by a female is as high as 27,000,000.

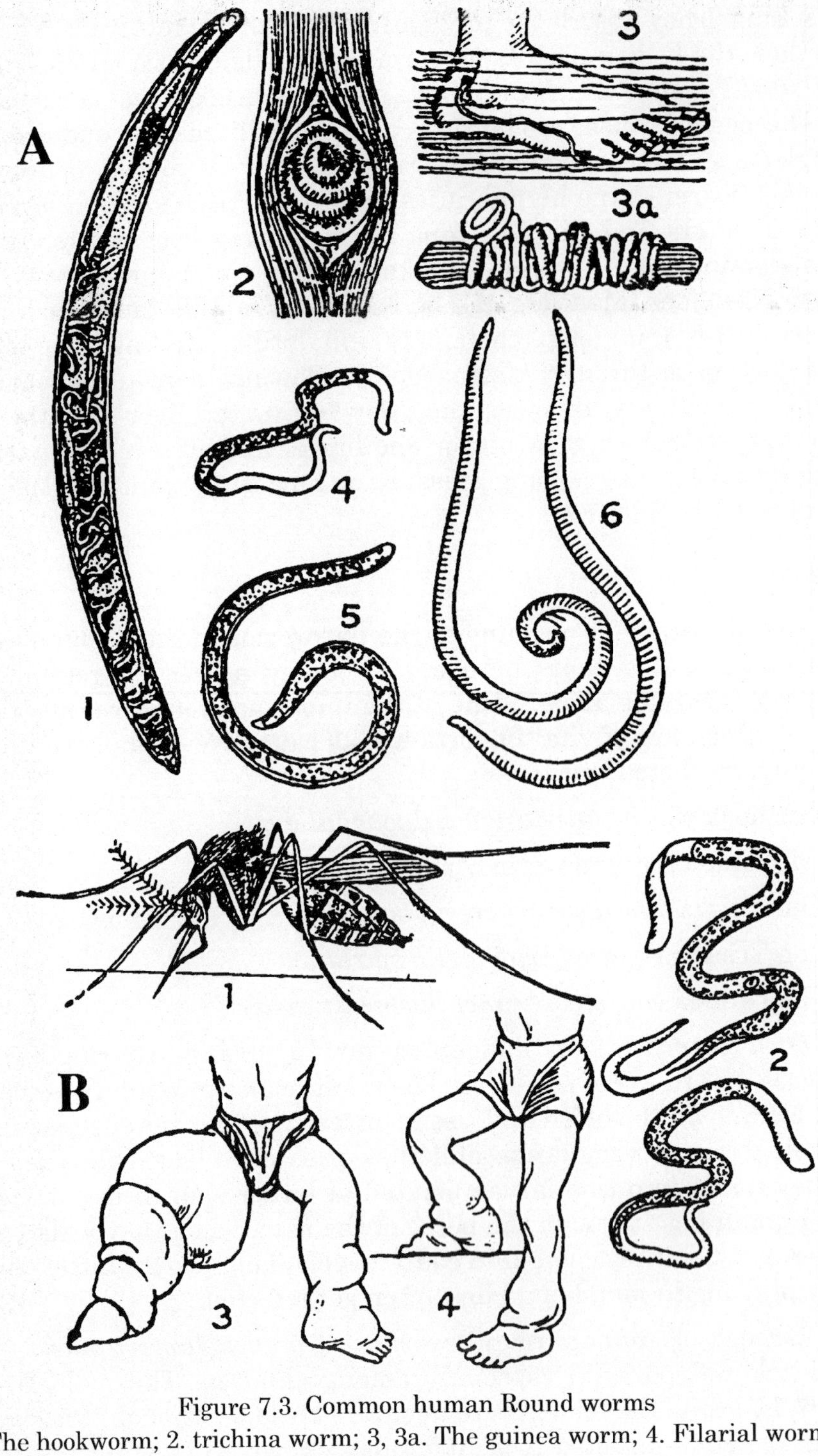

Figure 7.3. Common human Round worms

A. 1. The hookworm; 2. trichina worm; 3, 3a. The guinea worm; 4. Filarial worm; 5. Pin worm; 6. Round worm.

B. 1. Culex mosquito; 2. Filarial worm; 3, 4. Elephantiasis.

The causative organism for elephantiasis, the dreaded disease of the tropics is also a round worm called *Wuchereria bancrofti*. The mature worms reach almost 3" especially the females, and these are found in inextricable tangles in the lymphatic passages, blocking them. This results in the enlargement of the extremities like the hands and the legs, scrotum, etc. This disease is termed 'Filaria' or 'Filariasis' which is transmitted from man to man by the bite of a mosquito belonging to the *Culex* type. The mature females give birth to a number of young ones which are called *microfilariae* of a microscopic size, about 1/1000' long.

The trichina worm (*Trichinella spiralis*) is a microscopic thread worm often infecting man, through eating uncooked or improperly cocked pork. On reaching the intestine, the adult males produce countless young ones. These subsequently migrate to the muscles, where they roll themselves up into tiny balls. The toxic substances produced by myriads of the young in the human body may ultimately result in death. It is said that there are as many as 80,000 trichina worms in an ounce of cooked pork. Muscles which rich blood supply such as those of the limbs, ribs, diaphragm, tongue, etc. are favourite spots of invasion.

The guinea worm (*Dracunculus medinensis*) has been known to be a dangerous parasite of man, from ancient times and the technical name indicates the little dragon of 'Medina'. Today it is commonly found in India, Africa and S. America. The adult worms live in the tissues beneath the skin of the leg, ankle, foot and back. The females are large reaching, 3' in length, while the males are about 1-11/2 inches, particularly standing no comparison to the females.

The Phylum Acanthocephala includes the spiny head worms which form a compact group of endoparasitic worms in vertebrates and some invertebrates like molluscs, crustaceans and insects. Distribution of Acanthocephala is largely dependent upon the distribution of their hosts. The animal groups which harbour Acanthocephala at larval or adult stage are found in terrestrial and aquatic ecosystems. Some of these are highly pathogenic causing various diseases to swine, rodents and ducks.

Table: Species diversity of the major phyla of helminth parasites of animals

	Indian species	World species
Platyhelminthes	295	2,500
Monogenea	750	6,500
Digenia	530	4,500
Nematoda	250	20,000
Acanthocephala	229	800

Host-parasite relationship in the context of biodiversity and conservation

In many host-parasite relationships, the habits of the host must be taken into account in considering the distribution of the parasites. For example, dogs harbour the parasite hook worms *Dochnoides stenocephala* and *Ancylostoma caninum*, which are active throughout the year. The free development of eggs passed out through the faeces to the outside world, and the survival of ineffective larvae are possible only if the microenvironment at the ground level is conducive in the area where the host occurs. This may account for the restriction of the hookworms of dogs, *A. caninum* and *D. stenocephala* to specific areas, in spite of the wide range of distribution from the tropics to the arctic that the dogs enjoy. A number of environmental factors as well as the biology of the parasite, the host and their ecology in relation to the environment are totally needed for parasitism.

The host-parasite relationships attain a dynamic equilibrium when the association between the host and the parasite is a prolonged one with a large host population being infected every time and by the high specificity of the parasite. One of the important effects of parasitic mode of existence is the high specialization that the parasite evolves for its sustenance. This may be progressive as in *Taenia* spp., which have highly effective suckers and hooks for attachment, potential for prolific reproduction and adaptation for the procurement of liquid food; or regressive as in *Sacculina*, a crustacean parasite found on crabs, assuming a bizarre shape in the adult stage with just a sac full of eggs. Presumably, parasitism might have originated from commensalism or mutualism. Internal comensals or mutualists most likely would have evolved into parasites. Knowledge of host related parasite diversity and abundance is very essential and has obvious practical implications for the captive breeding and management of rare species of wild animals in view of the disease causing potential of animal parasites and disease transmitting potential of animal vectors. A high rate of contact between host and parasite encourages the spread of disease. Indirect effects of habitat destruction increases susceptibility to disease. *Ex situ* conservation programmes in particular, may pave the way to contact diseases from related species not usually encountered in their wild habitats and even from humans.

Parasites as historical probes for Biodiversity

Harold Manter (1966) stated that "parasites furnish information about present day habitats and ecology of their individual hosts. These same parasites also furnish information about the host and past biogeographical connections. They are the product of an immediate environment and a long ancestry reflecting association of millions of years". In other words, past is the key to the present and hence host-parasite relationships reveal the mystery behind historical biogeography, which includes the study of pattern and process in the distribution of organisms and historical ecology, which is concerned with macro evolutionary process in community development.

Parasites are indeed "biodiversity probes" which explain contemporary diversity and the historical development of community structure. The two significant phenomena of host-parasite evolution (coevolution, colonization or host-switching) and faunal distribution (endemism and centers of origin) are key stones to examining patterns of diversity. Most of the vertebrate and invertebrate fauna of the world is parasitized by flatworms, roundworms and acanthocephalans and these parasites provide new dimensions to understanding ecological interactions, patterns of distribution and the complex history of many geographical regions and biotas. Such information complements and increases the knowledge derived solely from the study of free living organisms on which parasites are dependent. Because of the unique insights gained from parasitiological studies about evolution of ecological interactions and community structure, parasitology is recognized as an integral constituent of biodiversity research.

Life cycles of parasitic helminths are usually complex. They are correlated with intricate food web and are dependent on specific ecological interactions and climatological conditions. Each species of parasite depends specifically on a species of intermediate host (an invertebrate) and a species of definitive host (a vertebrate). Thus, knowledge of the evolution of host-parasite assemblages provide insight into the history of ecological associations and community development. It also indicated that the well established trophic linkages constituting food webs evolved in the ancient geological past is retained to the present in an intact fashion.

An instance in point is the coevolution of the cestode flatworms of the genus *Livstowia* and the marsupials in the Neotropics and Australia. Life cycles of species of *Livstowia* are complex requiring both an arthropod intermediate host and a mammalian final host. Such a life cycle pattern has been evident before the geological separation of South America, Antarctica and Australia in the late Mesozoic to early Tertiary. Analysis of current distribution, life cycle pattern and phylogenetic relationship provides valuable information about ecological interactions over the geological distant past.

From a conservation point of view the evolutionary history of host-parasite assemblages indicates the age and duration of specific faunal associations, identification of regions of endemism and evolutionary "hotspots" and the historical structure of ecosystems and communities. These data will be helpful in prioritizing the phylogeneticlly significant lineages of taxa, biogeographically significant regions like centers of origin, radiation and endemism and unique historical ecological association for conservation.

Historical studies of host-parasite association also help in identifying potential climatological determinants of biodiversity. For example communities of sea birds and parasites in the Arctic region are strongly influenced by the environmental fluctuations that have taken place during the quaternary. Thus, the study of parasite communities in relation to respective hosts goes a long way in our understanding of the interactions of biotic communities, environment and climate. As a consequence, parasites constitute powerful tools and represent keystones to study the origin, maintenance and distribution of biological diversity.

Insect vectors of Human disease

Environmental changes that result from global and local processes have the same kinds of effects on vectors of human disease as on other insects. Both the distribution and abundance tend to change in ways that are not easy to predict. In particular, temperature changes associated with global warming could affect vector-borne pathogens in many ways, the most conspicuous of which occur by changing entire habitats and communities of organisms. Temperature controls the time taken for the parasite to mature in the vector. In other words it is the time an infected vector takes to become infectious. The major arthropod borne diseases and their vectors are included in the following Table 1.

Table 1. Insect orders (Class: Insecta), including those which have medical, economical or public health importance, with faunal wealth of species for each order (After Tyagi, 2004).

Subclass/ Order	Common names	Species number	Medical/veterinary/economic significance
Apterygota			
PROTURA	Proturans	90	Not known
THYSANURA	Silverfish	700	Economically important;
	Bristletails		Occasionally reported as biting humans
COLLEMBOLA	Springtails	2000	Important in forest ecology; also medically important
	Snow fleas		

Pterygota			
EPHEMEROPTERA	Mayflies	1500	Economically important
ODONATA	Dragonflies Damselflies	4500	Economically important; Predators on mosquitoes: rarely of medical and or veterinary importance
ISOPTERA	White ants	1800	Economically important: recently claimed to yield a new insecticide molecule
PLECOPTERA	Stoneflies	1500	Not known
EMBIOPTERA	Embiids, Webspinners	200	Not known
ORTHOPTERA	Grasshoppers, Locusts, Katydids, Crickets, Cockroaches, Praying mantids, Walking sticks	30000	Economically and medically important
DERMAPTERA	Earwigs	1100	Of some medical importance
PSOCOPTERA	Book lice, bark lice, psocids	500	Economically important; also of medical and veterinary importance
MALLOPHAGA	Bird lice, biting lice	2700	Economically important: also of medical and veterinary importance
ANOPLURA	True lice, sucking lice, crab lice	500	Medically important, transmit louse-borne typhus etc.
THYSANOPTERA	Thrips	6200	Economically important; rarely medically important
HETEROPTERA	True bugs		
HOMOPTERA	Aphids, Scale insects, Leafhoppers, Spittlebugs, Cicadas	60000	Economically important
COLEOPTERA	Beetles, weevils	280000	Economically important; and also medically important
STREPSIPTERA	Stylops, twisted wing insects	300	Not known
NEUROPTERA	Lacewings, Antlions, snake flies, Dobsonflies	2700	Not known
MECOPTERA	Scorpion flies	350	Of veterinary importance

TRICHOPTERA	Caddisflies, water moths	3600	Not known
LEPIDOPTERA	Moths, Butterflies, Skippers	125000	Economically important; some are medically important, too
HYMENOPTERA	Ants, bees, Wasps, sawflies, Woodwasps, Gallwasps, Ichneumons, Braconids, Chalcids	103000	Economically and medically important
DIPTERA	Flies, gnats, Mosquitoes, Midges, punkies	85000	Very important both medically and from veterinary standpoint, transmit several deadly diseases including malaria, dengue, JE etc.
SIPHONAPTERA	Fleas	1100	Medically important, transmit deadly diseases including plague etc.

Besides cultivated crops, stored products and other related damage caused to forest vegetation, insects are responsible for transmitting various kinds of diseases to man. Mosquitoes are well known vectors of four distinct human diseases. The adults of anopheline mosquitoes rest with proboscis and abdomen more or less parallel with surface. Dr. Ronald Ross in 1897 demonstrated that the *Anopheles* mosquito is the transmitter of malaria. The principal vectors in India are *Anopheles culicifacies*, *A. fluviatalis*, *A. varuna*, out of which the first three species are the major vectors in South India. *A. fluviatalis* is the main vector in hilly areas and *A. stephensi* in urban areas. Among culicine mosquitoes there are two groups viz., *Culex* group and *Aedes* group. In the *Culex* group eggs are laid in rafts and the siphon tube of the larvae is long. In *Aedes* group the eggs are deposited singly and the siphon tube is short and broad. *Culex quinquefasciatus*, found in urban areas in India, is the principal intermediate host of the filarial nematode, *Wuchereria bancrofti*, besides *Culex fatigans* (Fig. 7.4). The egg raft of *C. fatigans* is boat-shaped with about 200 eggs. Breeding is heavy in waters contaminated with sewage. Egg to adult takes 8-10 days. The microfilaria requires about 16-17 days in the mosquito before it could reach the inactive stage. Human infection occurs at the time of biting species of *Mansoniodes* are responsible for transmission of *Brugus malayi*. *Culex vishnui* is the vector for encephalitis, the reservoir for the virus being pigs.

Mosquitoes of the genus *Aedes* are responsible for transmission of the causative agents of two important diseases of man viz., dengue and yellow fever, *Aedes bancrofti* in Australia. The blood of a dengue patient is usually infective to the mosquito only during the first three days of the disease. The incubation period for the development of the virus in *A. aegypti* is 11-14 days and once infected, the infection lasts during the remainder of its life. *A. aegypti* is also the vector of the disease yellow fever. The infective material of yellow fever circulates in the blood of the patient for three days and during this time mosquitoes feeding on the patient become infective. It takes about 12 days before the mosquito is able to propagate the disease. Yellowing of the skin and eye is caused by the accumulation of bile pigments in the blood.

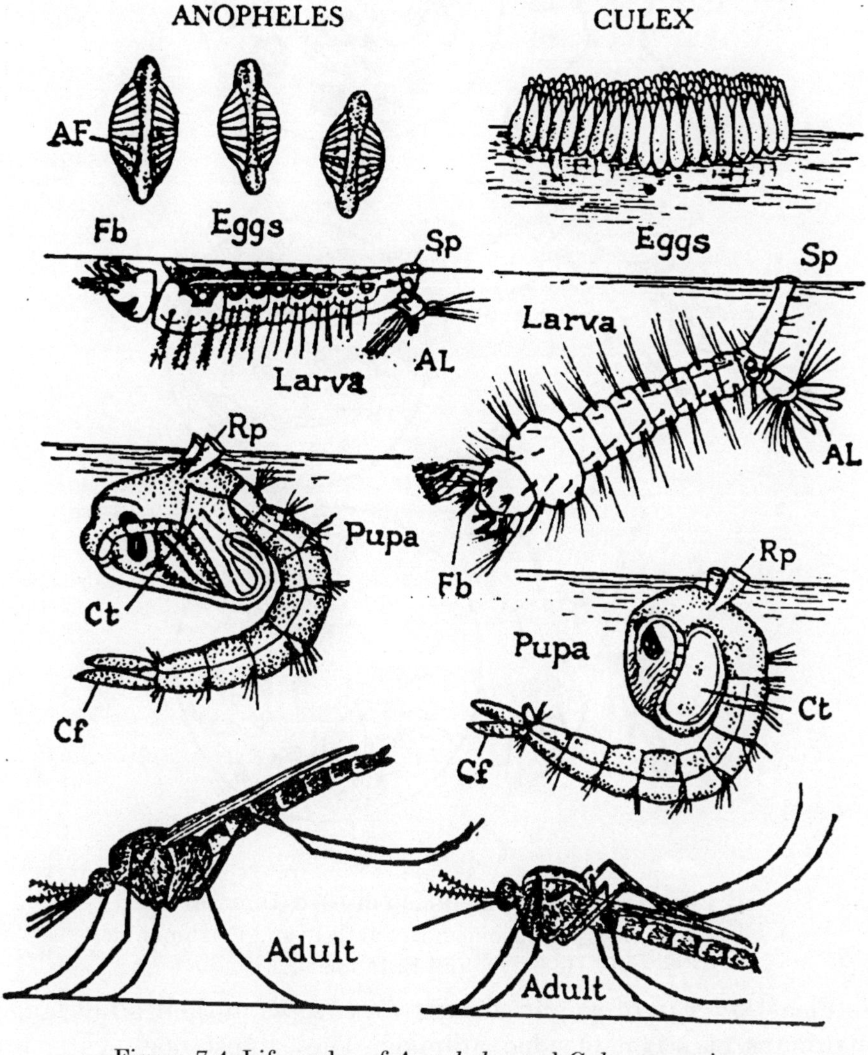

Figure 7.4. Life cycles of *Anopheles* and *Culex* mosquitoes

The sand flies (*Phlebotomus* spp.) are small with the body and wings thickly covered with hairs. The females of are blood-suckers and transmit the dreaded kala-azar disease caused by *Leishmania donovoni*, a flagellate protozoan, spread by *Phlebotomus argentipes* and *P. pappatasi* which transmit the pappatsi fever prevalent in various parts of India (Fig. 7.5). In Africa the nematode *Onchocerca* causes skin and eye disease and its vector is the black fly *Simulium*.

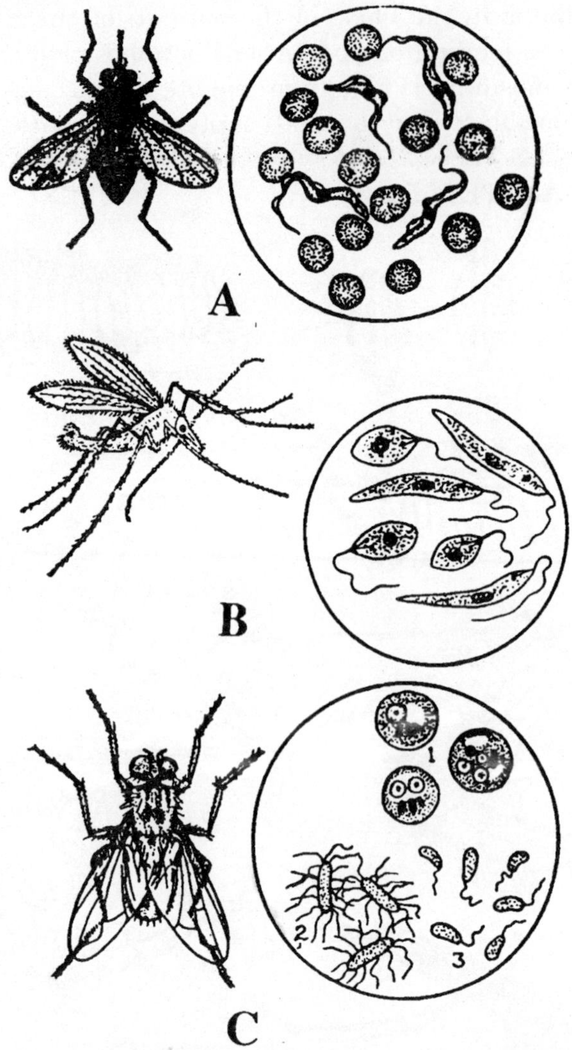

Figure 7.5. Vectors of human disease (Diptera)
A. Glossina and Trypanasomes; B. Sand fly and Leishmania;
C. House fly and Entamoeba.

The Fleas *Pulex irritans* and *Xenopsylla cheopis* in their adult stage live as ectoparasites of warm blooded animals. They infest dogs, cats, human beings etc. Their bite is not felt at once, but becomes increasingly irritating and sore for several days. Bubonic and septicaemic plague are the principal diseases carried by rat flea, *X. cheopis*. The causative organism of plague is the bacillus, *Pasteurella pestis*. *Xenopsylla cheopis* is widely distributed in the tropics and in India it is widely prevalent in the plains and also in the hilly regions. *X. brasiliensis* is another plague flea found in India and Africa. *Pulex irritans* is essentially a human flea and *Ctenocephalides canis* and *C. felis* are dog and cat fleas, respectively (Fig. 7.6).

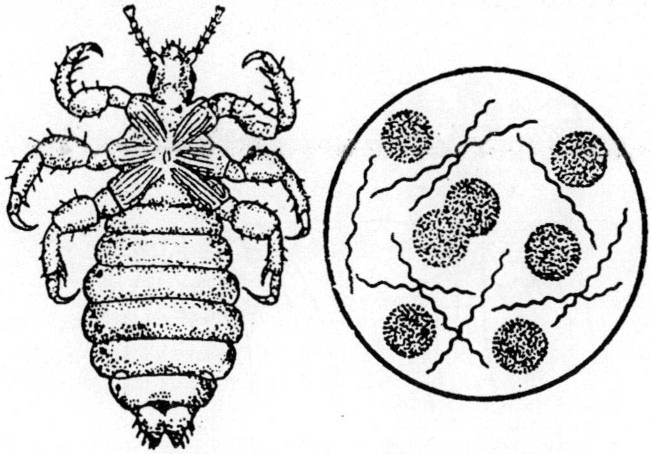

LOUSE AND SPIROCHAETES.

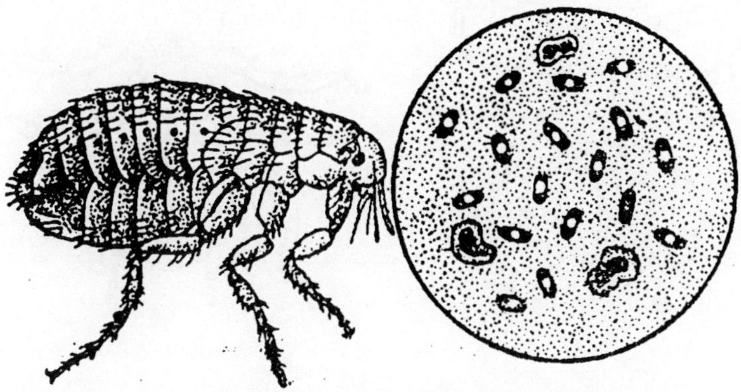

RAT-FLEA AND PLAGUE BACILLI:

Figure 7.6. Vectors of human disease (louse and fleas)

Ectoparasites associated with human beings are the sucking lice the commonest being the head louse, *Pediculus humanus capitis*, the body louse *Pediculus humanus humanus* and the pubic louse *Pediculus humanus pubis*. The body louse can bind up in epidemic proportion in no time and instances are on record, where one shirt taken from an infected individual having over 10,000 lice. Infestations cause intense itching and heavy infestation can cause discolouration of skin and called as the vagabond's disease. Lice can act as vectors of relapsing fever and typhus disease, caused by a bacteria-like organism *Rickettsia prowazeki*, which can develop only in the living cell.

Insects and other arthropods can cause direct injury by getting under the skin. The principal species involved in *dermatosis* are mites such as *Sarcoptes scabei*, the human itch mite which burrows into the skin, causing intense itching. Another example is the follicle mite *Demodex folliculorum*, seen inside their follicles (Fig. 7.7).

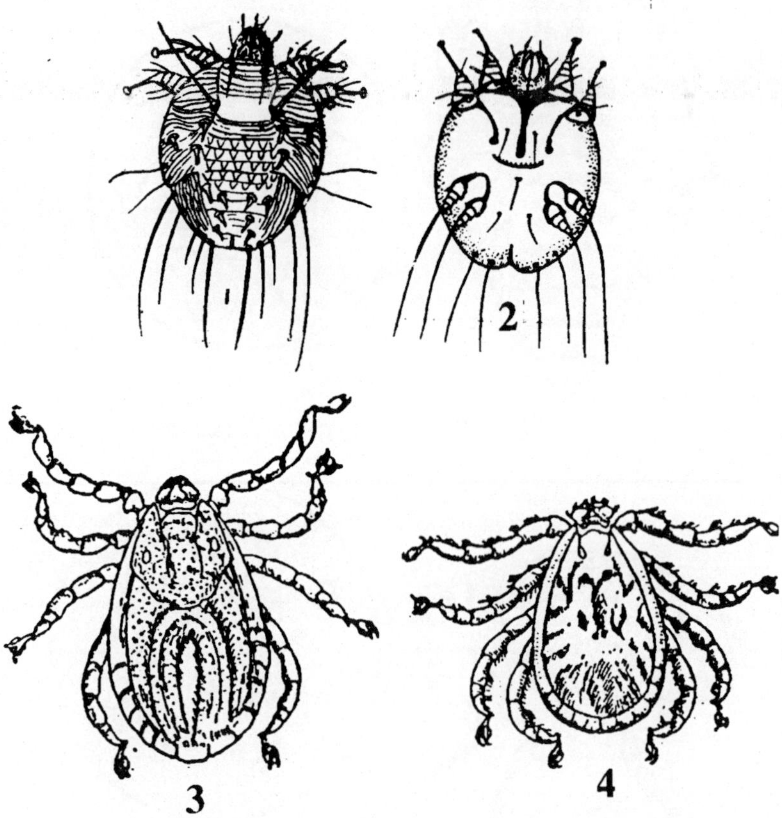

Figure 7.7. Vectors: 1,2. Itch mites; 3,4. Ticks.

Though the effects of global warming may be more significant than those of man-made environmental changes, yet elsewhere the effects of land-use changes like deforestation, irrigation and urbanization have resulted in profound effects on the distribution and scarcity of vector-borne diseases (Harrington and Stork, 1995).

CHAPTER - 8

INSECT BIODIVERSITY AND NATURE OF INSECT-PLANT INTERACTIONS

The evolutionary history of insects justifies their existence for millions of years and the passage of time has in no way diminished the wonders of insect life. Having their origin in the Carboniferous, they have shown evidence of extensive adaptive radiation in the Mesozoic, closely following the angiosperms, with further specialization in the Tertiary. Their close interactions with plants, animals and man are beyond comprehension and unquestionably they are man's closest and most relentless competitors, playing a vital role in the economy of nature. From the snow-clad heights of the Himalayas to the Tropical forests of the Western Ghats, from the deserts of Rajasthan to the rain forests of the North-East, they have exploited all kinds of environmental niches. Their numerical abundance is amazing in as much as it is inconceivable, with over an estimated ten million species with around 60,000 in this country. Many more await discovery, hopefully before many species are lost, especially in the Tropical forests, where increasing environmental depredation is rampant. Fields, forests and fresh water are not the only places where possible environmental depredation is rampant, but also the superficial layers of the soil, where one group of primitive insects, the Collembola or spring tails, whose numbers reach from a few thousand to 700,000 cm^2 in debris rich in organic matter, plays a very important role in the breakdown of organic matter in soil. The diversity of aquatic insects and their multifaceted adaptive mechanisms, as seen in dragonfly and damselfly nymphs, mayfly, stonefly and caddisfly larvae, besides whirligig beetles and mosquito larvae, are equally overwhelming. Groups of tiny flying insects constituting the aerial plankton are very common at heights up to 1500-2000 meters. Naturally, their feeding habits are diverse, involving phytophagy, carnivory, saprophagy, mycophagy, coprophagy and haematophagy. Besides, they also exploit the metabolic capabilities of microorganisms by harbouring symbiotic bacteria in their gut, not only for cellulose digestion, but also for supplementing the needed nutrients like amino acids, vitamins and sterols. Major wood feeding cerambycids like *Coelosterna scabator* harbour these bacteria in special diverticula of their hind gut. Insects like thrips, aphids, coccids, chalcids as well as others have the habit of initiating galls in specific plant hosts. The gall insects survive in the galls, exploiting them for nutritional purposes and for developing their progeny. The role of galls as nutritional sinks is a major feature presenting a highly evolved form of phytophagy.

In recent years the term coevolution has come to be used for the nature of the physical and chemical defenses of the plants against insects and other herbivores, and the adaptations involved therein. The diversity of plants would tend to increase insect diversity of plants. In other words coevolution is assumed to occur between two interacting species, each exerting selection processes affecting the other's gene pool. According to Ehrlich and Raven (1964), the diversification of modern insects is the result of step-wise patterns of coevolutionary changes superimposed on changing patterns of plant variation.

The coevolutionary interactions of insects with plants have formed the basis for increased understanding of the diverse role of insects, resulting in competitive interactions. Such interactions involve an integration of numerous complex, chemical and non-chemical factors. The defensive strategies of plants and the evolved ability of insects to overcome plant defenses have often been termed the 'evolutionary arms race'. Adaptive flexibility of insects to respond to ecological selection pressures is as important as the chemical composition of food plants. For instance, they are able to generate aerodynamic forces by executing angular oscillations, helping them to be airborne, and this development of flight has enabled the insects to be out of reach of their predators and other competitors. The mass flights of locusts should inspire scientific curiosity with their stratiform and towering swarms. Equally curious is the mass emergence and mass migration of the red hairy caterpillar *Amsacta albistriga* marching in millions from field to field of the groundnut crop. Seasonal, behavioural and physiological processes mediated by environmental factors govern insect-plant interactions. Their uncanny ability to choose feeding and oviposition sites naturally depend on their sensory capabilities, with information from the environment, such as colour, odour and taste, being encoded in the brain and subsequently decoded and integrated with the physiology of the insect to evoke a proper response. The focal point of such sensory capabilities lies in the antennae, the receptors on the sensillae aiding in the detection of plant odours.

Though often considered as loathesome objects in the farm and domestic fronts as agents of destruction and vectors of plant, animal and human disease, their beneficial role as producers of silk, honey, wax and lac, as pollinators and as biocontrol agents has come to be admired. But what has fascinated man is their unusual array of structural and behavioural diversity, their eggs, larvae and adults, providing unusual colour combinations which are a source of considerable aesthetic beauty and are much sought after. Studying caterpillars has been described as an adventure in wonderland. The social organization so well demonstrated by ants, termites, bees and wasps is equally amazing offering object lessons in the successful utilization of division of labour and producing engineering marvels in their nest-building operations. No doubt, insects form an excellent material also to be used as very good models to understand their ecobehavioural aspects at all levels of their existence.

Diversity of Insect Colouration

The diversity of insect colouration is unbelievably fantastic, ranging from metallic and iridescent colours, such as those of cantharid beetles, to individuals or combinations of white, orange, yellow and red of pierid butterflies or ommochrome pigments typical of nymphalid butterflies. Anthocyanin pigments absorbed from plants include scarlet, purple and blue colours. The green colour of grasshoppers, mantids, larval caterpillars and plant bugs is produced mostly by the biliveridin pigments, which have a blue green tint. The green is produced by an admixture of yellow, and yellow pigments are carotenoids as in, stick insects or xanthophyll as in long-horned grasshoppers. Several Lepidoptera show mottled olive green markings because of the juxtaposition of melanin (black) and xanthopterin (yellow) scales vividly coloured. The large saturnid moths *Antherea paphia, Samia cynthia ricini, Actias selene* as well as the largest of moths, *Atlas atlas,* are remarkable examples of such a colour pattern (Fig. 8.1). The golden, green bronzy or blue metallic lustre. The enamelled green of dragonflies is due to Tyndall blue with yellow. When carotenoids are combined with different proteins, varying colour combinations are produced. Anthocyanins have their best example in 'cochineal' from the cochineal insect. *Dactylopius tomentosus,* used to control prickly pear. The lac insect *Tachardia lacca* produces carmine.

Besides, glow worms and fireflies produce some of the most brilliant luminescence resulting in cold light or living light with the aid of light-producing organs, their continuous or flashing type of light emission being controlled by the nervous system. Light is produced when the enzyme luciferase acts on a substrate luciferin in the presence of oxygen and these beetles such as *Luciola gorhami* and *Lamprophorus tenebrosus* make use of the lights as mating signals.

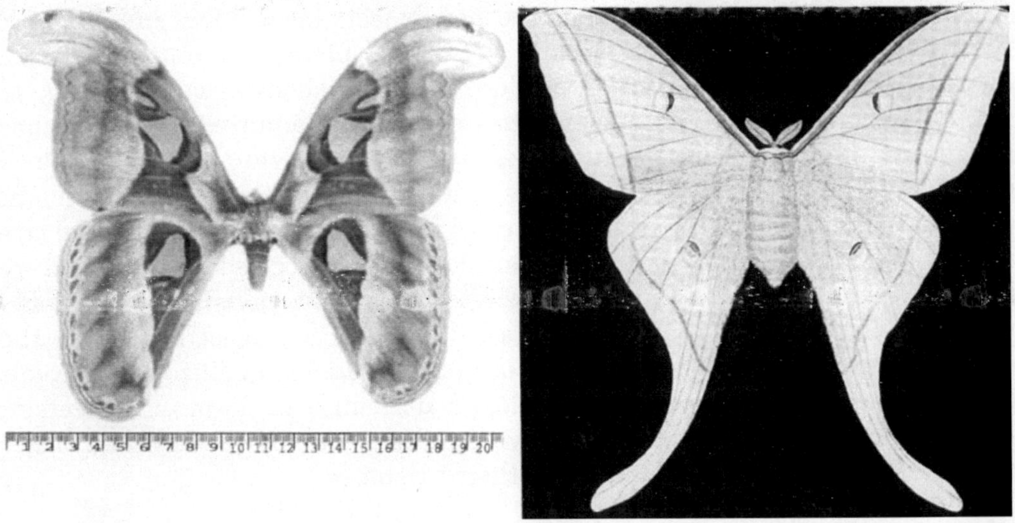

Figure 8.1. Some large-sized moths

Morphism and Functional Diversity

From the dimorphism of the sexes to complicated polymorphism or existence of a species in more than one form, insects provide meaningful examples of intraspecific diversity, the most notable of which is seen in social insects. Sexual dimorphism is evident generally in larger females and smaller males, but several instances of male specialization and sex-limited polymorphism in males are equally well known. Specialization in males is exemplified in lucanids or stag beetles (*Lucanus lunifer*), which have well-armed, enlarged mandibles for sexual combat. In the dung beetles or heliocoprids (*Heliocopris bucephalus*) and the rhinoceros (*Oryctes rhinoceros*) beetles, the enlarged horns are always outgrowths of the head and thorax. Earwigs also are good examples, with males having larger and stronger and armed forceps as in *Forcipula quadrispinosa* and *Labidura riparia*.

The locusts *Schistocerca gregaria* and *Locusta migratoria* provide striking examples of *phase dimorphism*, where there is a gregarious phase and a solitary phase, the former being energetic, with migratory properties and higher egg output. They show combinations of black, yellow and red colours, while in the solitary phase which lack these features they are greenish and more juvenile. Such group effects are equally common in Lepidoptera such as *Spodoptera exigua,* which are dull green on brown when solitary, assuming a velvety black colour when reared gregariously and also acquiring a migratory tendency.

Seasonal dimorphism occurs in many butterflies, aphids, leafhoppers and grasshoppers, where photoperiod plays a vital role, with short days and long days influencing the growth and development of these insects. The ability of these insects to respond in diverse ways to seasonally different day lengths suggests that they have a biological clock or a time-measuring system. Photoperiod involving as it does a cycle of a period of light and darkness provide environmental information in the form of time-spaced signals, resulting in diverse activity patterns. In aphids the viviparous parthenogenetic type of production to sexual, oviparous type is brought about by the short day lengths.

It is in social insects that we come across typical examples of polymorphism involving the queen, male workers, soldiers, as well as nasute soldiers among termites. Ants, besides all these diverse forms, show functional diversity to a maximum degree with fungus growers, seed collectors, thief ants, honey ants, dairy ants and slave-making ants.

Figure 8.2. 1. Termite mound: Queen termite (3) and workers (2)

Insect societies: their superorganism status

Insect societies, whether of ants, bees, wasps or termites, are wall organized in spite of their numbers. Marching colonies of ants several mutts long is a sight indeed. Sixty to eighty thousand worker bee; of four to five hundred thousand ants or termites form a single colony. Termite mounds reflect master design and engineering marvels in the control of ventilation, humidity and temperature. The queen termite is a phenomenal egg-laying machine, with its abdomen distended several hundred times the rest of the body (Fig. 8.2). In certain termites like *Bellicotermes* in Africa contain many million individuals which ceaselessly turn over the soil and vegetable debris. Paper wasps use weathered wood chewed to pulp mixed with saliva to form a durable 'paper mache'. Each insect society is regarded as a superorganism, with polymorphism and division of labour. Trophallaxis or food exchange plays a large part in their social set-up, especially of ants and termites and both have the habit of cultivating fungi in their nests. Behavioural changes are produced with the aid of social hormones or chemical substances which are passed on from one individual to another. The best example is the queen substance oxodecenoic acid secreted by the mandibular glands of the queen bee. Communication systems are typical of the bees, with the 'bee language' depicted by circular or figure of '8' and waggle dances translating the course to be in undertaken during forage.

Bees and ants find their way home by the polarized light of the sky. The eyes of insects are sensitive to natural phenomena that man is blind to, i.e. the polarized light of the daytime sky. The ability of the honey bees to navigate is also aided by the large number of visual units or ommatidia which can simultaneously scan many different parts of the celestial hemisphere.

Wasp studies have led to some of the most important discoveries in sociobiology such as nutritional control, caste and significance of dominance behaviour. Among social wasps a wide range of physiological behavioral adaptations have evolved to protect reproductive investment.

Chemical signals, the secret of behavioural diversity

Pheromones and allomones secreted by the exocrine glands into the environment act between different organisms, pheromones being intraspecific and allomones interspecific in their action. Hormones, on the other hand produce internal signals. Pheromones attract the sexes, aid in courtship, indicate danger, act as home trails and marker territories and control many intraspecific interactions. Releaser pheromones act rapidly to produce a behavioural response. Allomones repel predation, confuse prey and are more defensive. Pheromone signals occupy space and persist for some time. Sex pheromones attract sexes over long distance and pheromone detectors occur on the antennae of males. Interestingly enough, those occurring on the large feathery antennae of insects like the silk moth *Bombyx mori* are studded with 64,000 sensory hairs of which 80% are specialists responding to sex pheromones only. Many different species of moths emit pheromones simultaneously and somehow males distinguish conspecific female scents from others. For long-range orientation, anemotaxis functions, i.e. movement based on wind direction. Evidence of electromagnetic radiation (EMR) shows that during dusk, female moths become active, beat their wings continuously resulting in an increase of temperature over the ambient and at this temperature emit infrared radiation, which is picked up by male moths. In short, they operate as tiny electronic detectors of infrared; hence the concept of insect communication has come to be known as 'insect molecular bio-electronics'.

Visual cues also attract males and females in diurnal butterflies, males having a pair or large organs called hair pencils, which are tufts of fine hair-like processes. Males extrude the tuft of hair pencils from the tip of the abdomen for attraction and female lands in response. Mention may be made of pyrolizzidine alkaloids in plants, which are sequestered by the caterpillars and subsequently converted into dihydropyrolizzidine ketone which is a sex attractant. Attractants of moths are usually multicomponent blends emitted by the female and when pheromones concentration is above the threshold, males flying upwind until they come to the calling female. Most attractants are aliphatic alcohols, acetates or esters. Properties of each blend differ in signal molecules, distinguishing species. Several species of Lepidoptera show 'lek formation', consisting of groups of displaying males with extruded *coremata* attracting other males and females.

Nature of Insect-Plant Interactions

Insects and plants are contemporaneously evolving organisms with roots from the Devonian times and are involves in a nearly perfect symbiotic relationship, acquiring diverse adaptive strategies in terms of food resources for survival, growth, and reproduction. With their establishment from an aquatic to a terrestrial environment, plants became involved in competition, with a number of predators and were subject to multiple pressures, one of them being from insects. A similar development occurred in insects enabling individuals to feed on a variety of plants. In this competition, plants evolved resistance mechanisms by trying to utilize novel metabolic pathways producing compounds toxic to their predators. In course of time, insects also developed detoxifying mechanisms serving to overcome plant resistance. Naturally the numbers of such natural mechanisms were initially much restricted. But as insects evolved and developed more and more detoxifying mechanisms, plants began to respond through developing new deterrent substances.

The choice of the specific host plant by an insect besides depending upon the physiological conditions of both, also involves the balancing of external as well as internal excitary and inhibitory cues. Food plant selection therefore comprises a series of reactions whereby an insect rejects or accepts a plant. The maintenance of a stable insect/host relation and the evolution of new ones depend first on the ability of insects to find the plants with which their species is historically associated. Assessment of a plant mainly depends up on the sensory system conveying the requisite quantity of information about the environment. Thus the initial phase in host-plant relations, namely search and location also plays an important role in the catenary process of host selection by phytophagous insects. In addition, the ability of insects to perceive numerous chemical compounds in plants through the development of a highly developed chemosensory system appreciably influences the choice of the host plant. In recent years quite some efforts have been made to establish the causes for host selection behaviour, particularly the role of sensory physiology which has enabled a better understanding of the causes underlying the factors responsible for susceptibility as well as resistance. Therefore, to be more successful, environmental versality on the part of insects appears to be an essential aspect aiding in the development of structural, physiological, and behavioural characteristics which enable them to behave differently in different habitats, at different periods of their development and under varying conditions of environment.

Feeding Guilds

Feeding guilds in phytophagous insects have been recognized, many of them being xylem or phloem sappers, stalk borers, leaf miners, fruit and seed feeders and so on. Host plant specificity is a sequel to specialization of food requirements and naturally diverse patterns of feeding behaviour exist, enabling different phytophagous insects to respond to such changes within the plant as ageing, water stress, concentration of nutrients and the like. As plants evolved, feeding preferences changed to accommodate plant changes. Insects become tropically associated with several plants so as to make possible the recognition of *eurytopic* and *euryphagic* or *stenophagic* insects which are more restricted in their food range.

Based on the degree of feeding specialization, oligo- and polyphagous species are often recognized. While a polyphagous species is able to occupy widely different ecological niches and geographical ranges, it also tends to prevent over-exploitation of preferred plant species. On the other hand, monophagous species tend towards greater specialization in restricting to their hosts. The rate of development of an insect on different plants, often tends to vary and host alteration in most cases has also the advantage of escape from egg predation and possible inter-specific competition. In some cases host alteration serves only to provide water, sugars and other compounds just to keep the insects alive, but not sustain reproduction. Many insects pass the dry season as active adults in reproductive diapause, so that in the dry season the degree of host preference decreases. But such alternate hosts in which they do not reproduce, appear to influence the fitness of the insect when it locates the primary hosts. The ultimate choice of a plant by an insect is not directed by its nutrient characters alone, but it is the stimulating effect of the nutrient compounds such as sucrose, glucose, amino acids, sterols and other compounds that enhance their feeding responses. At the same time insects also tend to influence plant population levels, the degree of diet specialties being an important factor governing host suitability which depends on several synecological factors.

Table 1. Ranking insect-feeding guilds (with examples) according to their potential impact on host plant (after Mattson *et al.*, 1988)

Herbivore guild	Potential impact on host growth and reproduction
Gall formers, leaves	Least
Folivores-end-of-season/prior-year leaves (Diprionidae)	↓
Folivores-beginning of the season/current-year leaves (Lymantriidae, Tortricidae)	
Sap feeders (Aphididae, Coccidae)	
Phloem/sapwood/pith borers (Cerambycidae, Curculionidae)	
Root sap feeders (Cicadidae)	
Folivores-middle/late season (Arctiidae, Lymantriidae, Notodontidae	
Folivores-both current and prior-year leaves (Diprionidae, Lymantriidae)	
Sap feeders-stem, phloem/xylum (Aphidiae, Coccidae, Diaspididae)	
Phloem/cambium/sapwood borers (Buprestidae, Curculionidae, Cerambycidae)	Most

In the determination of insect numbers and life history strategies, it is the host plant which plays a critical role, since the habitat or area accessible to the feeding stages of insects are involved. Further, the plant species being a heterogenous entity with its microenvironment changing in time and space, conditions for growth and other factors also change. Fresh leaves after pruning become more acceptable, resulting in increased preferences in relation to egg production and duration of life cycles.

Insects and Plant Galls

Another important area where plants become adapted to the attack by insects relate to the biology of gall formation. Plant galls are abnormal growths produced by plants in response to the feeding or oviposition stimulus of insects resulting in excessive cell multiplication and growth. Galls are endowed with more favourable conditions of microclimate, providing an optimal environment for rapid reproduction and abundant food for the duration of the season. Initiation and exploitation of plant tissues leading to the formation of plant galls by insects is considered as a highly developed form of phytophagy. The evolution of the gall forming capacity also depends on the habits and behaviour of the cecidozoans and this is very obvious among chalacids, gall midges, some gall wasps, ants and tephritids living in higher plants, where not only sufficient nutritive substances are available, but also undifferen-tiated meristematic tissues, ideal for the development of a gall. Gall insect- host plant association seems to exemplify an advanced/ specialized level of 'trophic strategy'. This is evident from the organization of a nutritive zone in galls, an adaptation *par excellence.* The gall forming or the cecidogenous habit is evident in many species of thrips, aphids, aleyrodids, coccids, psyllids among Homoptera. Among the Diptera, gall makers include the agromyzids, tephritids, cecidomycids, while the Hymenoptera include chalcids, cynipids and saw-flies. While these are the recognized gall makers, a few curculionid beetles and some moths are also known to form galls. Some of the common gall type include:-

Filz galls are characterized by the presence of hairs or hairy outgrowths of the epidermis. Most filz galls are induced on the underside of the leaf blade and in this type of gall the inducer is situated externally. Many mite galls are of the filz gall type. *Pit galls* and *blister galls* are simple galls characterized by a slight arching of the leaf blade. The epidermis sometimes splits to form the blister. The simplest of these galls are hardly visible and are called "rudimentary galls". *Pouch galls* develop first as a bulging of the leaf blade. They are often induced on young leaves by the very first stimulation of the gall insect. As the name implies, pouch galls are typically hollow. There is usually an opening to the outside referred to as an ostiole. Pouch galls are provoked mostly by Acarina, aphids, cecidomyiids, and psyllids. *Roll* and *fold galls* may present different degrees of complexity. Depending on the stage of swelling and the differentiation of the lamina, numerous types of roll and fold galls are produced. In leaf roll galls the blade rolls upward from the leaf margin toward the midrib and there is swelling of the rolled-up part of the leaf blade. In fold galls the unfolding of the leaf rudiments inside the developing bud is suppressed and affected. *Bud* and *rosette galls* range in complexity from simple enlargements of buds, the so-called "big buds", to solid or hollow swellings with spiny or scaly surfaces representing the bud leaves. Most bud galls are usually globose, but some are flask shaped. Some galls are fleshy swellings of all leaf rudiments. Rosette galls, resemble a miniature cabbage or pine cone. In complex bud galls there is a pronounced swelling and fusion of the parts giving rise to a fleshy or woody mass (Fig. 8.3).

Figure 8.3. Insect galls

Ant-Plant Mutualism

The mutualisms between ants and plants emerge as solutions to problems in the lives of the participants. For example, without being able to move, plants must escape herbivores and find suitable sites for their seeds to grow. Interaction with animals can provide these services, repelling herbivores or dispersing seeds. Plants can provide nest sites and/or food for ants and so, when ants defend these resources from other animals, they are also defending their host plants. Seeds or pollen carried by animals, intentionally or unintentionally, will be transported to other places.

At the same time, ants use plants to solve their problems. Ant colonies need a protected nest for their vulnerable eggs, larvae, and pupae. Nest sites are usually in soil, but in many ecosystems good nest sites are in short supply, so that plants that provide nest sites are almost always occupied by ants. Since ant foragers always search for food, they gather it wherever food is available from plants. Thus, many different ant-plant mutualisms have evolved as ants and plants use each other to increase their own fitness.

Ants form an important group of plant mutualists because of their abundance and distribution. Plants have mutualistic interactions with many kinds of animals, but ants are probably more numerous and widespread than any other single type of animal. Ants inhabit virtually every ecosystem in the world and within ecosystems, the total number of ants can be very great, with literally thousands of workers per square meter. The social structure of ant colonies enhances their effectiveness as mutualists: A single ant colony may have foragers searching hundreds of places at once, or send many different individuals to the same place. As a result, ant-plant mutualisms include relationships in which plants are defended from herbivores or have their seeds and pollen transported. In return, plants provide ants with shelter and/or food.

Most ant species consume some seeds, but harvester ants eat predominantly seeds and are major herbivores in some areas. Ants of the New World tropical tribe Attini consume plants indirectly by feeding them to a fungus and eating the resulting fungal mycelia. Both seed-harvesting and fungus-gardening ants maintain colonies with hundreds of thousands of workers, and greatly affect the composition of the plant community wherever they occur.

Indirect interactions also occur between plants and ants. For example, ants may indirectly injure plants by protecting plant-feeding insects such as aphids and membracids. Ants tending membracids have in some cases been shown to provide indirect benefit to plants by chasing off herbivores which cause more damage to the plant than do the membracids. Ant nest-building activity turns the soil, aerating it, and differentially benefiting particular plant species.

Myrmecophilous Plants

Ant loving plants are well known, being protected from herbivore attacks, reflecting a more complex relationship, besides symbiosis.

Myrmecophytes as these are called, the plants are classified into *myrmecotrophic* (providing food), *myrmecodomic* (providing a dwelling) and *myrmecoxemic* (both together) (Jolivet, 1992). More than 25 families, 40 genera and 270 species of plants are known to be myrmecophilic. These are large hollow spines which shelter ants in Acacia; some rubiaceous plants like Hydrophytes have the stem inflated within where ants live. The inflated hollow tubercles, pitchered or modified leaves, inflated and modified internodes occur in species of Acacia. Ants play an important role for these plants as defenders and scavengers, nutrition providers, seed disseminators and gall generators (Fig. 8.4).

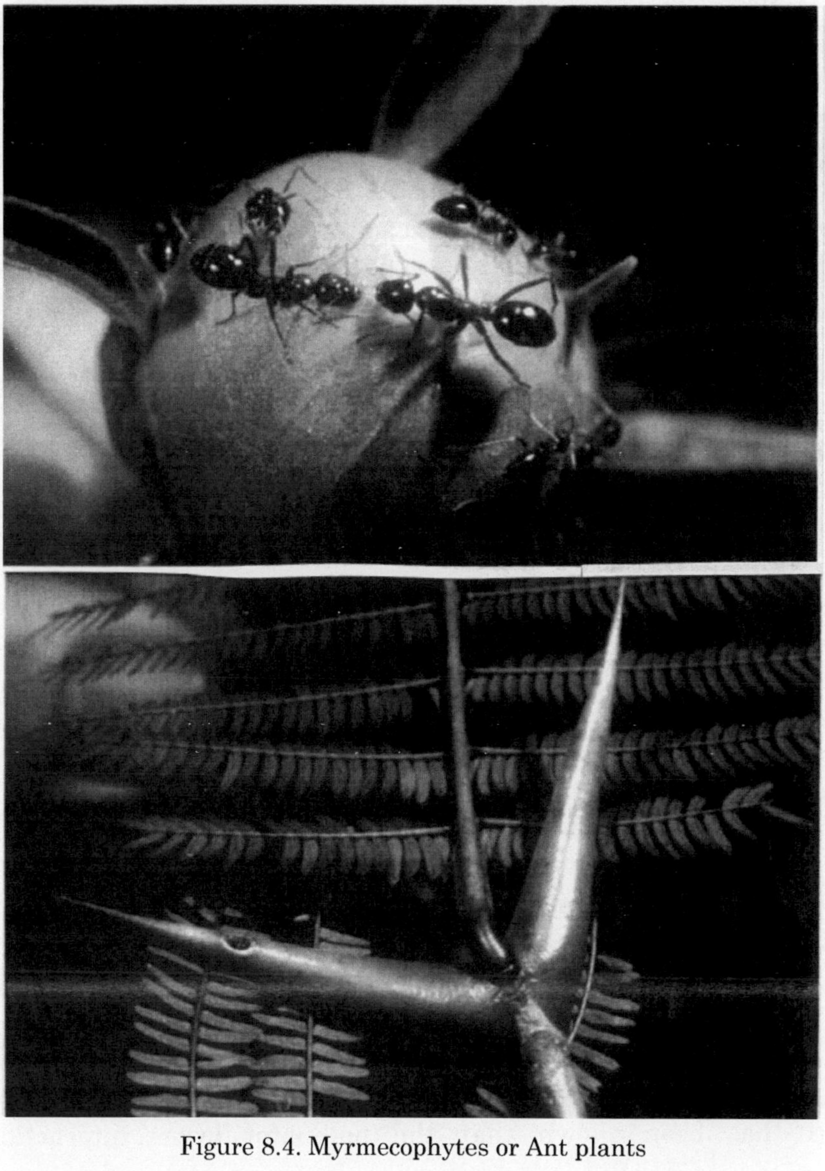

Figure 8.4. Myrmecophytes or Ant plants

Insect Pollinator Diversity

The flower as the functional reproductive unit provides a vital link in plant breeding systems with the pollinating insects forming, an integral part of the evolving biological system, the pollinator and the floral organization functioning as a closely-knit entity. Plant-pollinator interactions have been widely recognized since the cretaceous times and the significance of these interactions relate to the fact that they have contributed considerably to each other's evolution. This is clearly evident from the rapid adaptation of plants to particular microhabitats resulting in the proliferation of angiosperm species in the cretaceous period, due to pollinating insects. The absence of efficient pollinators like the bees and the butterflies before the appearance of angiosperms and their subsequent appearance with further specialization of plant families, their flowers providing landing platforms for such pollinators exemplify the coevolutionary trends involved. The basic trends in floral evolution have been reduction in size of floral component, concealment of nectaries, fusion of floral parts, aggregation of flowers, inflorescence and development of attractants to pollinators. Availability of pollen and nectar, of perfumes, of protective as well as visual sites, and of sexual attraction are among the principal attractants responsible for establishing blossom-pollinator relationship. The role of floral odors in pollination is well known and bees and other pollinators are known to be attracted to specific chemical compounds produced by extra-floral structures of flowers themselves. Bees have been known to be trained to locate forage by associating it with particular floral odours, the osmophores functioning to select the potential pollinators.

Besides providing proteins and amino acids in quantities, pollen also provides a range of essential and quasi-essential amino acids, while nectar from the nectaries represents an aqueous solution of energy producing sugars. Nectar has an array of chemicals-amino acids, proteins, lipids and sugars as energy producers; ascorbic acid, an antioxidant and nutritional source alkaloids which may be toxic to flower visitors, protective in function. Flower-insect coadaptation therefore is a continuing process, since the dynamics of pollen transfer mechanisms are closely related to the behavioural characters of the pollinators. Pollinators have, therefore, become an integral part of intrafloral ecology involving energetics.

While bees, butterflies and moths are among the major pollinators, other animals like birds and bats also play a very useful role in pollination (Fig. 8.5). Normally the pollinators may need more than one plant for feeding and the plants are also exposed to more than one pollinator during the blossoming period. It has been shown that the pollinator-flower interactions are essentially a community based one, since a constant energy flow through the community has to be maintained at a minimum level to maintain the pollinator community. An optimum pollinator would be one which would contact the anthers and stigma more rapidly among plants, remaining continually with the flowers or the plants even in the presence of other flowering plants. It often happens that nature favours flowers secreting smaller quantities of nectar per flowering plant, so as to enable visits of insects to other plants. The diversity of flowering pollinators naturally results in the diversity of feeding preferences, so that categorisation into 'specialist' and 'generalist' feeders is common. Among bees for instance, there are species that collect pollen from a few related flowers in the same genus or even in a single species. These are termed as *oligolectic* or *monolectic*. There are others which collect pollen from many different or unrelated kinds of flowers and these are broadly designed *polylectic*. Examples of moths and flowers evolving through reciprocal selection is seen in the long tongued moth selecting long-tubed flowers, which in turn select for long- tongued moths adapted for reaching nectar.

Feeding specialization varies from site to site or within sites during different periods depending on the resource availability or degree of competition involved, so that what is normally oligolectic can turn into a polylectic species. According to the nature of foraging activities, the following categories are recognised, (a) indiscriminate foragers which have a few behavioural or morphological links with flowers, (b) facultative specialists which encounter a dense cluster of flowers, (c) obligatory specialists capable of identifying rewarding flowers with nectar while switching to more profitable flowers and (d) efficient foragers which are predisposed for visits to a particular flowering species and remains with it. If flowering is confined to the entire foraging lifetime of the insect, then selection will favour particular flowers so that synchronization occurs in nectar and pollen production. Generally speaking, it is the amount of nectar and pollen reward per flower and must sufficiently justify the visit, but it should not be excessive enough to prevent or limit successive flower visits. Based on this assumption, flower visits by insects have been explained on the basis of the 'optimum foraging theory', defined as that which yields the greatest net energy and nutrient gains per unit foraging time. Naturally the energy budget of the pollinator will vary from species to species governed mostly by the body weight, temperature etc. Plants control the calorific reward for particular insects, since each has a different energy requirement. Insects requiring higher levels of energy levels therefore not utilise plants with low nectar content.

Table. Floral syndromes of some animal pollinated plants (after Baker and Hurd, 1968; Faegri and van der Pijl, 1979).

Syndrome	Pollinator(s)	Anthesis	colour(s)
Cantharophily	Beetles	Day and night	Variable, usually dull
Sapromyophily	Carrion and Dung-flies	Day and night	Purple-brown or greenish
Myophily	Syrphid Bee flies	Day and night	Variable
Melittophily	Bees	Day and night or diurnal	Variable but non pure red
Sphingophily	Hawkmoths	Usually nocturnal	White, pale or green
Phalaenophily	Small moths	Usually nocturnal	White, pale or green
Psychrophily	Butterflies	Day and night or diurnal	Bright red, yellow, or blue
Ornithophilly	Birds	Diurnal	Bright red
Chiropterophily	Bats	Nocturnal	Dull white or green

The flowering constancy or patterns of foraging of pollinators are high when it is restricted to a single species promoting co-evolution between the insect and the plant, as it is typical of some bees. Knowledge of the foraging dynamics is therefore essential for a better appreciation of the efficiency of pollinating systems in plants. Though flowering systems are well known to anthecologists, many more instances are needed to be able to properly interpret the dynamics of such systems.

Phalaenophily or moth pollinated flowers include the non-hovering noctuids, zygaenids, and pyralid moths as well as the swift-flying hawk moths. Members of the families Aristolochiaceae and Asclepidaccae, have flowers smelling like excreta, luring dung and carrion flies away from their usual food sources to the flowers - a phenomenon termed *Sapromyophily*. The pollinator-flower interactions are discussed in this chapter based on unique interacting systems such as, Thrips-Compositae system; Bee/Butterfly-flower interactions, and caprification phenomenon.

Bird Pollinators

Major bird pollinators include the New World hummingbirds (Trochilidae), the Old World sunbirds (Nectariniidae) and honeyeaters (Meliphagidae), and representatives of a few other taxa. Bird flowers are brightly coloured, often red or yellow, and lack odor and a landing platform. These features correspond to the visual orientation of these pollinators and the fact that they do not generally perch on the flowers. Nectar is abundant and deeply held, accommodating the high energy requirements and long bills and/or tongues of birds. Red colouration has been suggested to be especially appropriate for attracting birds exclusively, because it is conspicuous to them, but indistinguishable from a grey background to many insects.

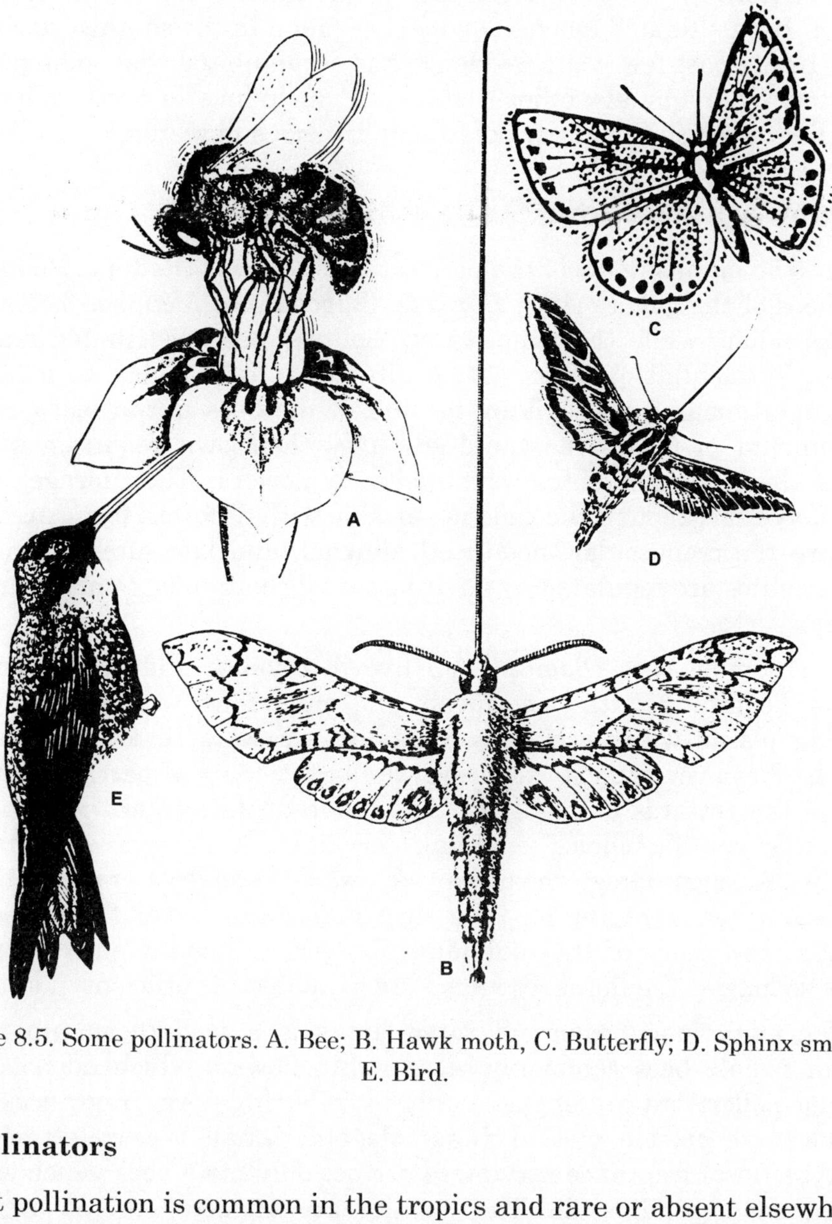

Figure 8.5. Some pollinators. A. Bee; B. Hawk moth, C. Butterfly; D. Sphinx smoth; E. Bird.

Bat Pollinators

Bat pollination is common in the tropics and rare or absent elsewhere. It is associated with a distinctive suite of characteristics including flowers that are often borne in unusual positions: on long, hanging stalks, outside the foliage, or springing directly from large stems (cauliflory). These type of arrangements facilitate visitation by large, awkward visitors. Flowers open in the evening or at night and frequently last for just one night. They are whitish or dull-coloured, but emit a strong odor and are large and open, providing easy access. Pollen and nectar are produced in large quantities, the latter sometimes amounting to several milliliters per flower. Some tropical genera exhibiting bat pollination include *Bauhinia, Ceiba, Adansonia,* and *Mucuna.* Bat-pollinated agaves *(Agave)* are found in the southwestern United States. In the past few years, evidence has accumulated that some plants are pollinated by mammals other than bats, including lemurs, rodents, and marsupials, such as the honey possum and various phalangers.

Bees-Flower Interactions

Bees accomplish 80% of the pollination by insects (Reddi & Reddi, 1984), social bees of the genera *Apis, Bombus, Trigona and Melipona* belong to the family Apidiae, while the majority are solitary bees distributed among the families, Megachilidae, Halictidae, Andrenidae, Xylocopidae, to mention the more important ones. A striking feature of the bees is that they require a large amount of food, even hundred times their own requirement and to achieve this they make frequent visits to flowers. They forage normally during day time, though the time is variable with different bees. Accordingly, there are the crepuscular, nocturnal, diurnal, and late afternoon bees and their activities are regulated by environmental conditions, temperature being the major factor.

The bumble bees (*Bombus* sp.) live in colonies and large numbers of foragers beginning with the queen and numerous workers exploit wild flowering plants with each forager having access to several plant species. They also have the capacity to associate floral colours of particular plants in terms of the rewards available so that different bees from the same colony have their specific choice enabling foraging on flowers with increased resources. As such though the colony as a whole receives a broad diet brought by different bees foraging on their own major and minor flora species, yet from the view point of the individual forager, it 'majors' on a single plant species so long as the floral resources are available in sufficient quantities.

The euglossine bee-orchid interactions is a well-known phenomenon wherein female bees commonly visit orchid flowers which do not produce nectar or pollen, but are engaged only in collecting floral fragrances such as benzyl acetate, cineole, methyl cinnamole, etc. Female bees collect nectar and pollen the floral fragrance serving to attract only male bees which swarm in numbers on a small scale. The females are attracted to the swarm and mating takes place. Males pollinate orchid flowers of the genus *Ophrys* by attempting to copulate with flowers which resemble female insects. What is of interest is that the plant species adapted to these bees have evolved different types of floral fragrances to attract different pollinators and simultaneously the presence of several natural plant chemicals influence the development of the bee species. While the bees are dependent on the flowers in view of their specificity to a fragrance, even minor changes in their composition tends to disrupt their attraction to flowers.

Butterfly-Flower-Association

In contrast to the social hymenopterans, lepidopterans do not nurse their brood and hence the food collected by the adults are utilised for their own consumption. Butterflies and moths restrict themselves almost entirely to nectar as their food. This adaptation has reduced their moth organs to long thin suctorial tubes that can be inserted into the long corolla tubes. These special flowers usually visited by butterflies possess long tubes or spurs inaccessible to short tongued insects. The tendency of a pollinator to move from a plant to one of its near neighbours is evident in butterflies and moths. The flower nectars on which the adult lepidoptera feed, contain significant concentrations of a wide range of aminoacids, which contribute a lot to its nutritional value.

Butterfly proboscis is clearly adapted for reaching nectar at the base of long tubed flowers and different species vary greatly in their proboscis length. Butterflies are diurnal and some of them are only capable of using shallow blossoms like those of *Melaleuca sieberi* (Myrtaceae).

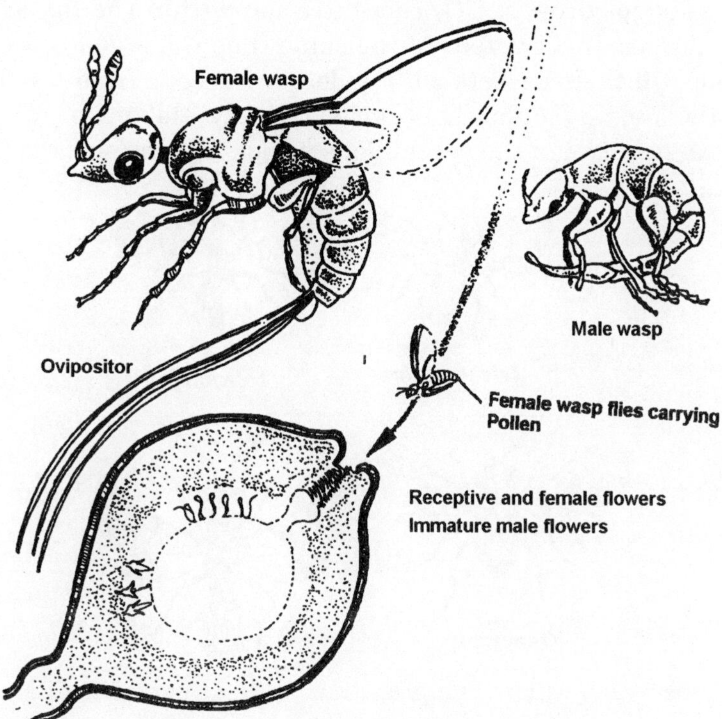

Figure 8.6. Caprification, Fig pollination

Caprification

One of the best examples of interdependence between plants and insects is that of figs and fig wasps. These are established pollen vectors of various *Ficus* species which are mostly host specific. Caprification increased harvest of figs through cross pollination by wasps is an established phenomenon in the pollination of *Ficus* species. Wasps possess specific pollen containers that have a role in the transport of pollen. Pollination by wasps involves two phases, i.e. (i) loading of containers with pollen by female wasps from their native figs before they emerge (male phase); and (ii) unloading them after entering young receptive figs (female phase). The pollination act itself takes place at the end of each oviposition, before the wasp withdraws its ovipositor from the pistil. In *Ficus sycomerus* detailed studies on the pollination act by the wasp *Ceratosolen arabicus* revealed the existence of unique thoracic pollen pockets. Pollination is directly effected by the wasp during egg laying. Recent investigations on the fig physiology, has highlighted the micro environmental conditions within the fig which play a decisive role in the regulation of the activities of wasps. Increased fig respiration results in CO_2 accumulation in the internal atmosphere to as much as 10% and under these conditions the males of the pollinators *Blastophaga quadraticeps* fertilize the females which are still inactive. Tunnelling of the syconial wall by the males results in an abrupt drop in CO_2 concentration within the fig, as a result of which the males are inactivated. At the same time the females become active, leave the galls, fill their pockets with pollen and escape from the fig cavity for pollinating the next syconia. Thus an extremely elaborate and precise co-adaptation exists between the sycophilous wasps and the syconia (Fig. 8.6).

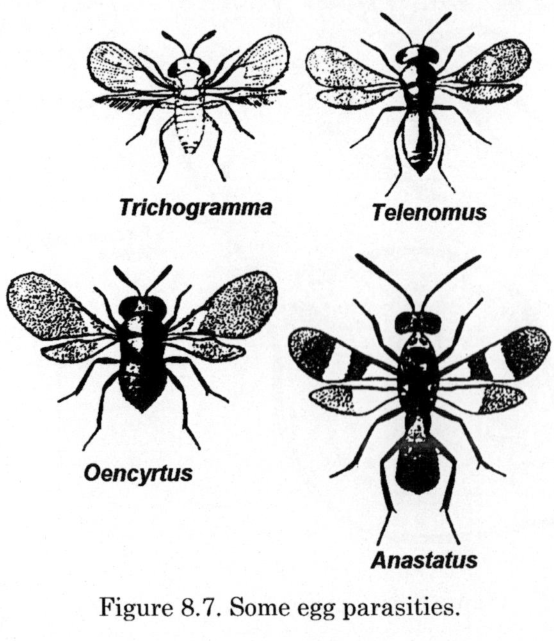

Figure 8.7. Some egg parasities.

Biological control and Biodiversity

The remarkable diversity of highly specific natural enemy species – parasites and predators is made use of to control pests of importance in agriculture, through increasing the diversity and abundance of natural enemy species in crops, as well as the diversity of natural enemies in exotic situations. As control agents for the pests which are also native to another country, it has become possible to agument natural control resources. About 1/10 of all species on earth are insect parasitoids which are of great value in biological control. Host size influences the species composition in egg parasitoid assemblages of Lepidoptera. For e.g. species of *Anastatus* are found only in assemblages of large-sized hosts, since they cannot parasitize small sized hosts. Egg parasitoids other than *Trichogramma* which contains the smallest species are more restricted. *Telenomus* another egg parasite, which cannot parasitize such small hosts as can *Trichogramma*, is less polyphagous (Fig. 8.7). For egg parasitoids, hosts size is a prime determinant of their species richness and habitat specification of each parasitoid species largely determines the species composition of each assemblage (Hirose, 1994).

A number of strategies for increasing natural enemy diversity are known. A more diverse habitat will support a greater range of natural enemy species. Where there is no natural vegetations, flowers or shelters around a field crop would be of use, because greater the area surrounding natural vegetation, the greater the species diversity not only of plants, but also of natural enemies. Encouraging natural vegetation around crops would be of use, but the strategy of polyculture or growing crops together, could create a greater diversity (Letourneau, 1987).

CHAPTER - 9

BIODIVERSITY IN AGROECOSYSTEMS

Agroecosystems comprise the total complex of organisms in the crop area, together with the overall conditioning environment as modified by diverse human activities. It is a complex of mutually interacting network of food chains. The major components include the crop plants, soil substrate and its essential biota, the chemical and physical environment and energy inputs. Weeds, plant pathogens and insect pests are dominant components of biodiversity that is important to agriculture and help ensure sustainability, stability and productivity of systems regardless of the complexity of the ecosystem. Such diversity tends to lead to a more efficient use of natural resources through providing greater opportunities for nutrient recycling. Within any particular species, genetic diversity is most important, since it is the variation that enables that species to adapt to the new ecosystem and environment. Genetic diversity within a cultivated crop species at the field or farm level helps to diminish the risk of losses through diseases and pests. Genetic diversity provides the reservoir of genes for future crop improvement. Modern agroecosystems depend on the constant new identification of genes that occur in genetically diverse crops and such useful genetic diversity can also occur in wild relatives of crop plants and this diversity must be conserved and appropriately used for improving crop performances. Thus diversity is important to agriculture at all levels and in all agroecosystems.

Conservation of genetic resources is essential so that informal breeders have access to the genes and gene complexes for current and future crop improvement. Besides this will enable farmers to continue to select and modify their crop in response to changing environments. Wild non domesticated relatives of crops frequently provide useful sources of genes. For

Figure 9.1. Sugarcane top borer. 1-9. Various stages in the life cycle.

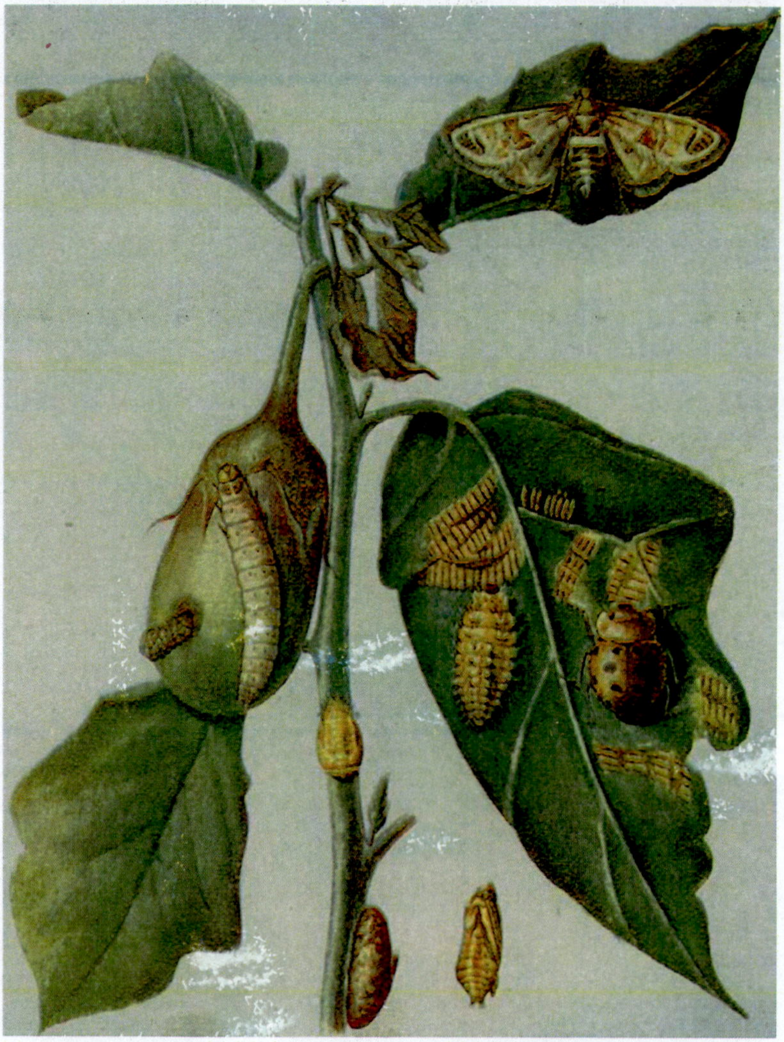

Figure 9.2. Brinjal borer and coccinellids.

example, a wild rice *Oryza nivara* was used to introduce resistance to the grassy stunt virus. In Africa and India, cassava yields increased up to 18 times after introducing genes from wild cassava, conferring disease resistance. In today's scenario, new DNA sequences can be created or introduced into crop species. With modern genetic engineering techniques potentially useful genes could be transformed between crops and induced to express themselves. These new genes become integrated into the plant genome. As plant breeders continue to effectively use genetic resources, their conservation becomes essential, so as to have continued access, to genes and gene complexes for crop imporvement. Biotic diversity at levels ranging from genetic diversity among populations to landscape diversity is critical to the maintenance of natural agroecosystems.

Pests and disease tend to cause damage to cropping systems decreasing productivity. Some are key pests in each agroecosystem, like for instance rice, cotton, sugarcane, etc. These key pests are permanently occurring, resistant species whose level of damage and abundance are important criteria. When the infestation is serious it would be necessary to hold populations at low levels. Other pests could be occasional, potential or migratory pests which are described as minor or sporadic pests. Occasional or sporadic pests cause economic damage only in certain places or at certain times. Potential pests are considered minor pests, since in certain situations they become problematic. Migratory pests are usually non existent in agroecosystems, but appear all of a sudden in a crop area for their breeding places, causing heavy damage and leaving subsequently. The locusts and army worms are good examples of migratory pests. Southwood (1976) adopted the term r-pests and k-pests and intermediate pests. Classical pests like locusts, aphids and other major pests belong to the r-category, while k-pests have low rates of potential increase, more specialized food habits and seldom become pests of economic importance. Southwood points out that there is a continuum between the two extremes of r and k strategies and most of the crop pests are intermediate.

Biodiversity contributes in a variety of ways towards ecosystem functioning, such as production, decomposition, nutrient cycling. The cultivation of multiple species necessarily entails biodiversity aspects. Choices of cropping combinations are a biodiversity strategy influencing production. There are many agricultural practices and designs with the potential for enhancing functional biodiversity, through applying the best management practices for enhancing or rejuvenating the kind of biodiversity which provides ecological services like biological control as well as nutrient cycles in soil conservation. A healthy soil with direct effects of plant quality on herbivores and a rich natural enemy biodiversity, are positive strategies.

Profile of Insect Damage

The nature of the damage inflicted by most pests is related to their feeding habits which are essentially sucking and biting and chewing. Direct injuries result form larvae scraping off tissues, or roll leaf blades or mining through leaves. Larvae of *Cnaphalacoris medinalis* on paddy roll the leaf length-wise, while *Sylepta derogata* infesting cotton and lady's finger plant cut themselves about the middle of leaf and then roll it. The larvae may form a protective covering by cutting away leaves as in *Nymphula depunctalis* in paddy. The citrus leaf miner *Phylonestis citrella*, the tea leaf miner, *Gracilaria theivora* and the cowpea leaf miner, *Cosmopteryx* sp. mine through the leaves forming irregular or zigzag passages. *Scirpophaga incertulas* causes chaffy earheads or dead hearts, while *S. xanthogastrella* the sugarcane top borer feeds from the top destroying the crop (Fig. 9.1).

Many insects feed internally within the stem causing drying up or dying of branches, causing dead hearts. The commonly known bark-eating caterpillar such as *Inderbela theivora* on tea, and *Inderbela quadrinotata* on Moringa, guava etc., *Chilo partellus* on wheat and maize causing drying up of infested branches are good examples. Internal damage can be caused by gall formers like *Orseola oryzae* on paddy causing the silver shoot. Swelling of branches by feeding inside is typical of the cotton stem weevil *Pempherulus affinis*. The adults of the Rhinoceros beetles *Oryctes rhinoceros* bore into the unopen fronds of coconut palms causing death of the plants. Cutworms like *Agrotis* sp. cause damage during night by clipping off various plants. The larvae of *Mythimna seperata* remain hidden during the day in cracks of soil or below bushy shrubs and cause severe damage to sugarcane after sunset. Similarly the grubs of cockchafer beetles feed on the roots of various plants including groundnut and sugarcane.

Thrips and hemipteran bugs cause serious damage. The gregarious habits of aphids feeding on the lower sides of leaves cause curling of leaves. Jassids or green flies feed on the lower surfaces of leaves, causing yellowing, curling down of leaves and in severe cases, leaf shedding as on lady's finger by *Amrasca biguttula*. On beans *Aphis craccivora* causes twisting of the stem. Infestations by sucking insects also cause malformation and galls on different plant parts. Some insects like aphids, whiteflies and jassids act as vectors of virus disease.

Reproductive parts of plants are damaged by internal feeders causing drying, decay, destroying and shedding of fruits. The larvae of brinjal borer, *Leucinodes orbonalis* destroy fruits (Fig. 9.2), while larvae of *Helicoverpa armigera* cause empty bored legumes. External feeders cause skeletonising of leaves by feeding caterpillar larvae of *Amsacta albistriga* and *Spodoptera litura*. Scraping of leaves in a circular fashion is done by larvae of *Plutella xylostella* on cruciferous plants. Complete destruction by larvae feeding only on the midrib veins is done by the larvae of *Spodoptera litura, Achaea janata* in castor.

Some Important insect pests in agroecosystems (Figs. 9.3, 9.4 & 9.5)

Lepidoptera

Aproarema modicella, the leaf miner of groundnut

Amsacta albistriga, the red hairy caterpillar on groundnut

Opisina arenosella, the black headed caterpillar of coconut

Sylepta derogata, the cotton leaf roller

Helicoverpa armigera, the cotton bollworm

Platyedra gossypiella, the pink bollworm

Earias insulana, the spotted bollworm

Earias vittella, the spotted bollworm

Chilo infuscatellus, the shoot borer of sugarcane

Scirpophaga nivella, the rice stem borer

Chilo polychrysa, dark headed borer

Spodoptera mauritia, the swarming caterpillar or Armyworm

Nymphula depunctalis, the rice case worm

Figure 9.3. Major pests of Agriculture - Adults

Hemiptera

Nephotettix nigropictus, the green rice leaf hopper
Nephotettix virescens, the green rice leaf hopper
Nilaparvata lugens, the brown plant hopper
Leptocorisa acuta, the rice earhead bug
Aphis craccivora, the lablab aphid
Dysdercus cingulatus, the red cotton bug
Oxycaraenus hyalipennis, the dusky cotton bug
Stephanitis typicus, lace wing bug
Aleurolobus barodensis, the whitefly of sugarcane
Pyrilla perpusilla, sugarcane leaf hopper
Melanaspis glomeratus, sugarcane scale

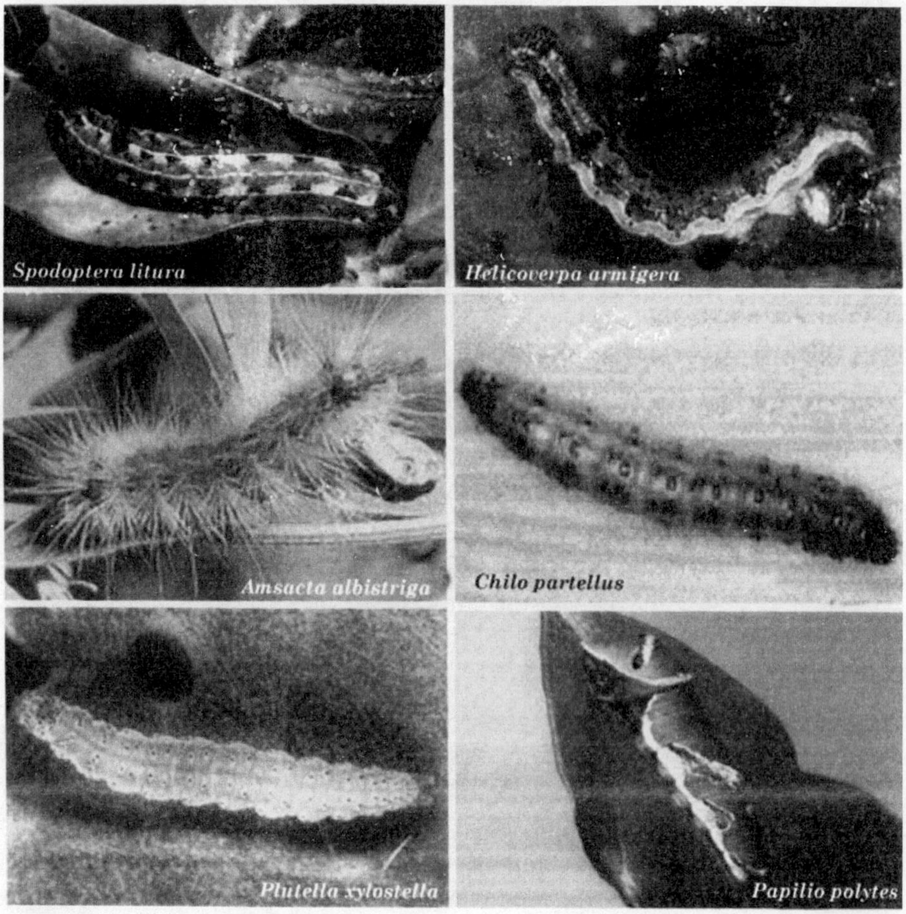

Figure 9.4. Some major caterpillar pests

Coleoptera

Echinocnemis oryzae, rice root weevil

Rhynchophorus ferrugineus, red palm weevil

Oryctes rhinoceros, the rhinoceros beetle

Pempherulus affinis, the cotton stem weevil

Alcidodes affaber, the shoot weevil

Figure 9.5. Some agriculturally important pests

It may not be out of place to mention some of the major pests in plantation crops. Insects and mite pests are major limiting factors in achieving high productivity in plantation crops. *Helopeltis theivora* the tea mosquito, the shot hole borer, *Xylosandrus compactus* in coffee, *Sciothrips cardamomi* of cardamom, the coffee berry borer *Hypothenemus hampei, Oryctes rhinoceros* and the dreaded eriophyid mite *Aceria guerreronis* threatening coconut plantations to mention only some of the major pests.

Biological Control of Pests

Major parasitoids belong to four distantly related genera *Trichogramma, Telenomus, Oenocyrtus* and *Anastatus* all parasitizing lepidopteran eggs. The smallest is *Trichogramma* which can attack host eggs of all sizes and the largest is *Anastatus* can parasitize hosts of the same size. The other two are intermediate. Among the parasites used in the biological control of many of these pests are the *ichneumonids* with a long ovipositor and *braconids*, which are endoparasites feeding on the internal tissues of the host insects (e.g. *Cotesia flavipes, Bracon brevicornis*). Chalcids are primary or secondary parasites of lepidopteran pests are of considerable importance in the biological control of a wide variety of insects (Fig. 9.6).

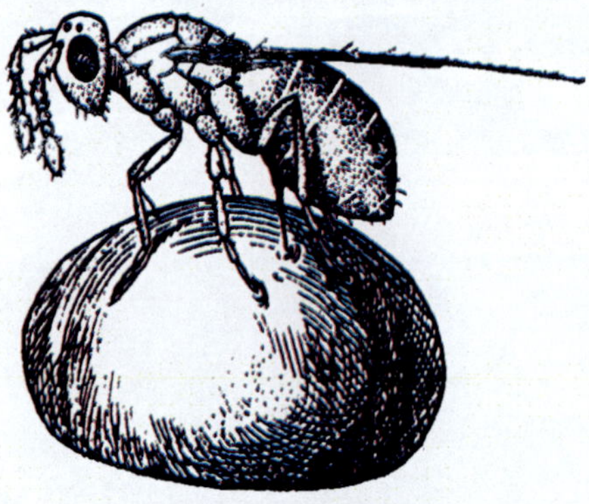

Egg parasite

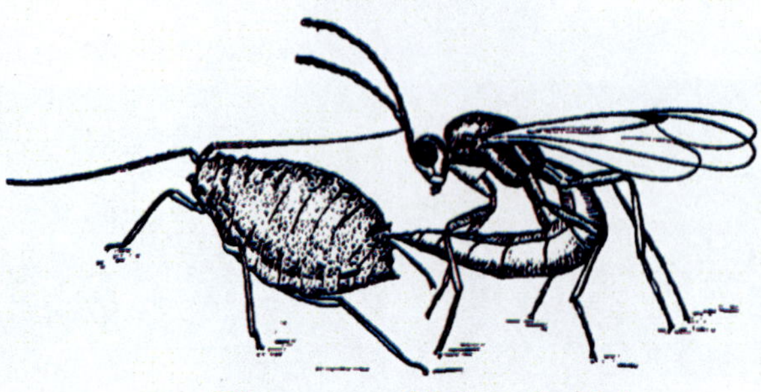

Wasp parasitizing Aphid

Figure 9.6. Egg and Larval parasitoids

Other biological control agents are arachnids such as spiders, mites and nematodes. Phytoseid mites such as *Typhlodromus*, *Phytoseius* are common examples. Among nematodes are mermithids, parasitic in ants and mosquitoes. Many aquatic insects such as notonectids, nepids, belostomatids are predatory on mosquito larvae. Syrphid flies and lacewings, tiger beetles, praying mantids, lady bird beetles, stink bugs or pentatomids, reduviids are well known predators. The use of pathogenic fungi such as *Beauvaria bassiana* against caterpillars and *Metarrhizium anisopliae* against beetles and viruses such as *baculovirusus* and *granulosis* viruses against lepidopteran insects are common. Parasitic nematodes like *Neoaplectana* are used against beetles. Several predatory insects are known to feed on eggs, larvae and adults of several insects' anthocorids, reduviids (Hemiptera) besides several Coleoptera and Diptera. While conservation and enhancement of natural enemies indigenous to a locality has been in vogue, the importation and colonization of natural enemies of indigenous and exotic pests are equally prevalent. Mass culture and release of natural enemies has been a well known practice.

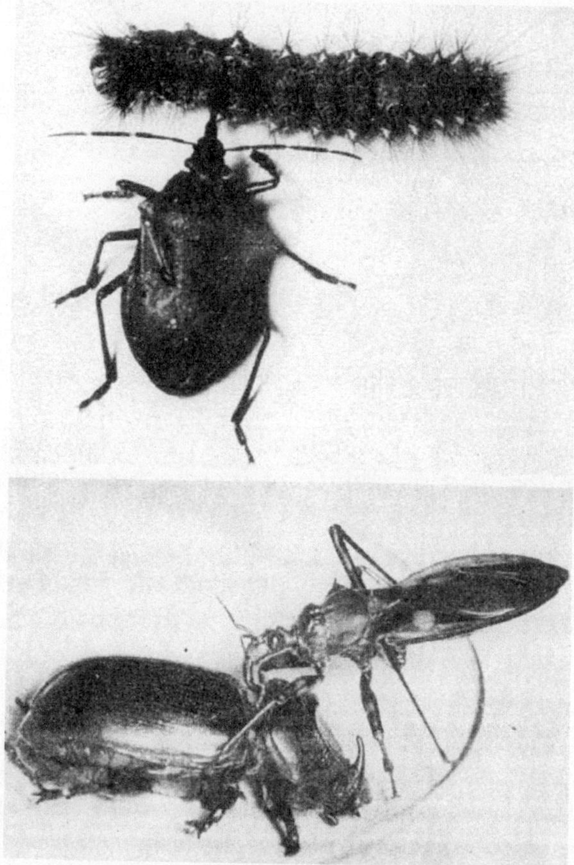

Figure 9.7. Reduviid bug (bottom) and pentatomid bugs (top) predatory on insects

Some of the more important predators and their prey are indicated below (Fig. 9.7).

Menochilus sexmaculatus *Coccinella septumpunctata* *Rodolia cardinalis*	Coccinellid or ladybird beetles
Rhynocoris marginatus *Carayonocoris indicus* *Orius insidiosus* *Orius maxidentex* *Montandoniola moraguesi*	Reduviid bugs or assassin bugs
Cantheconidia fuscellata *Tropidonotus guttatipennis*	Pentatomid bug
Leptomantis indica *Gonyptera punctata*	Mantids
Episyrphus batteatus *Sphaerophoria indica* *Metasyrphus enfrater*	Syrphids
Amblyseius tetranychivorus *Tetranychus urticae*	Mites

Insect vectors of Plant disease

Plant pathogens are transmitted by insects, besides others like nematodes and mites, the transmitting insects being called vectors, which belong mostly to the Hemiptera or bugs, particularly the aphids, leaf hoppers, whiteflies, mealy bugs, besides thrips and a few beetles. The common virus diseases are vein decaying diseases, yellow mosaic, leaf curl, banana bunchy top, cardamom mosaic etc. Generally one type of virus disease is transmitted by one group of insects, the mosaics by aphids, and leaf curls by whiteflies.

For transmission of viruses, activity of the insect vectors is more important rather than their number. In the case of aphids, it is the activity and number of migrant insects that is important in the efficiency of virus transmission, rather the number of apterous individuals, which, of course, are important in respect of their injury to the crop.

Some of the insect vectors and the viruses transmitted by them are (Fig. 9.8)

Insect vector	Virus
Pentalonia nigrinervosa	Banana bunchy top; cardamom mosaic (Marble or 'Katte' disease
Aphis gossypii	Papaya mosaic, Cucumber mosaic, Chilli mosaic
Myzus persicae	Cucumber mosaic
Aphis craccivora	Cow pea mosaic, papaya mosaic
Rhopalosiphum maidis	Sugarcane mosaic
Bemisia tabaci	Bhendi yellow vein mosaic (or vein-clearing disease), Dolichos Yellow mosaic, Tobacco leaf curl, Tomato leaf curl, Papaya leaf curl,
Thrips tabaci	Tomato spotted wilt.

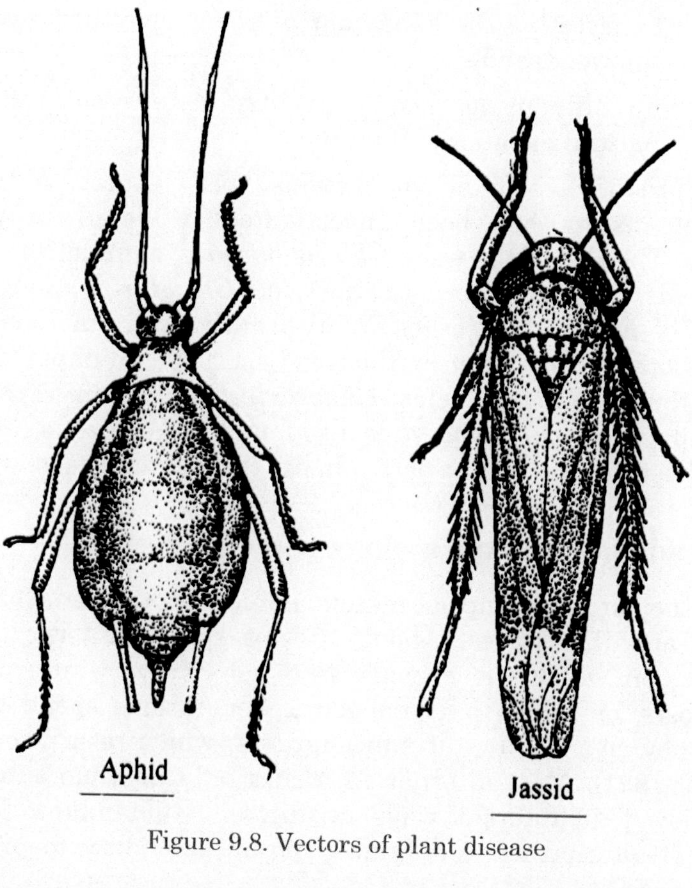

Figure 9.8. Vectors of plant disease

Role of Weeds in Agroecosystems

The ability of weeds to harbour insects is well-known and Van Emden (1981) has cited about 442 examples of weeds or wild plants as hosts of crop pests or disease organisms. Weeds are so prevalent that they are an important component of agroecosystems.The fact that many pest species colonize on weeds, and breed successfully reveals that they can migrate to the crop and inflict severe damage at any time. This is more so when the growing period of crop plants synchronize with that of the weeds. Instances where there are definite host-plant successions between crops and weeds are also on record. Thus a knowledge of the nature of colonization of insect pests and their population build-up on crop and weed hosts become essential to understand the interactions between insects, crops and weeds.

Weeds may influence the insect and other invertebrate pests of crops in many ways (Way and Cammel, 1981).

A. by acting as alternate or alternative hosts of the pests;

B. by affecting the biology of insect pests;

C. by affecting the visual responses of flying insects;

D. by affecting the behaviour of pests once they have contacted the weed; and

E. by influencing pest mortality, as a result of harbouring natural enemies of insects.

The nature of weed-crop interactions with respect to the seasonal fluctuation of insects has been investigated in detail in polyphagous Thysanoptera. The role of the weed *Chloris barbata,* abundant in *Pennisetum typhoides* fields, as an alternate host of *Chirothrips mexicanus.* The inflorescence of *Echinochloa crusgalli* also harbours numerous adults and larvae of *Haplothrips ganglbaueri*, the weed acting as an important alternate host for the thrips in paddy fields. *Caliothrips indicus* an important thrips pest of groundnut, *Arachis hypogea* is also found to colonize in large numbers on the weed host *Achyranthes aspera,* almost throughout the year.

Effects of wild plants on the biology of Insect Pests

Among the various groups of insects, aphids are very much renowned for their complicated life cycles. About 10% of species alternate between a primary host, on which the sexuals give rise to overwintering eggs, and secondary hosts, on which parthenogentic generations spend the summer. Eventually they give rise to autumn migrants, which return to the primary host (Eastop, 1981). Most alternation occurs only in some subfamilies and may be achieved in different ways in different subfamilies. For instance, alate males of alternating Aphidinae fly from secondary to primary hosts, while in other groups sexuparae fly from the secondary hosts to produce apterous males on the primary host. While most alternating aphids return annually to the primary host, almost regardless of conditions, some individuals may remain on the secondary host in the autumn reproducing parthenogenetically. The nature and extent of survival, growth and reproduction of insects on crop and weed hosts ensures not only the suitability of the crop for the colonizing species of insects, but also reveals the ability of different weed species to harbour different species of insects.

Weeds Harbouring Natural Enemies of Pests

Many instances can be cited with regard to natural enemies with free living stages that feed on the pollen and nectar of wild plants and with predatory/parasitic stages that feed on prey or hosts on wild plants (Ananthakrishnan, 1986). Flowers of many weed plants provide sumptuous food for natural enemies, for example, adults of some Syrphidae and parasitic Hymenoptera. Many of the predators can be increased through seasonal manipulation of selected weeds and effectively used as natural enemies of important crop pests. Early stages of crop growth, the presence of specific weeds may act as a reinforcement for natural enemies of crop pests and thus sustain the population of predators. Altieri *et al.* (1977) reported a chemical interaction between the leaf hopper *Empoasca* and two grass weeds, *Eleusine indica* and *Leptochloa filiformis* when grown on the boundaries of the bean field, exerted a repellent effect decreasing its colonization efficiency. Weeds also play a role in increasing the antiherbivore chemical defense of certain crops.

Destruction of Lantana bushes with a tortricid moth, an agromyzid fly, a trypetid fly and a tingid bug is a typical example of insect control of weeds, not to mention of the control of the prickly pear (*Opuntia* sp.) by a moth *Cactoblastis cactorum* and the control of the famous water hyacinth *Eichornia crassipes* by the weevil *Neochetina eichorniae* are well known instances of the biological control of weeds.

Effects of agriculture on species diversity

The distribution of plants and animals has been greatly influenced by the spread of agriculture and Cox and Atkins (1979) have compared it to major geographical forces such as the continental drift in determining the distribution of species. In particular, the introductions of exotic crops can bring with them associated insect species, besides other antagonistic interactions such as that of crops and weeds, plant pathogens and insects and plants. Weeds are so prevalent that they are an important component of agroecoystems. Many pest species colonize on weeds and breed successfully, migrating to the crop inflicting severe damage. A knowledge of the nature of colonization of insect pests and their population build-up on crop and weed pests are essential aspects in relation to the interaction between insects, crops and weeds. Wild populations typically have more genetic variation than their domestic counterparts in agroecosystems. This genetic diversity permits adaptation to a range of environmental conditions, so that in the event of changes this genetic diversity confers an ability to adapt and persist. The following table provides a comparison of the diverse characteristics of organisms in natural and agroecosystems (Abrahamson, 1989).

Table 1. Comparison of ecological characteristics of organisms and ecological processes in natural and agricultural ecosystems (After Abrahamson, 1989)

Characteristics	Agricultural Ecosystems	Natural Ecosystems
Life history and genetic characteristics of plants and animals:		
Life-history strategies	r-selected	k-selected
Life cycles	Short (annual)	Long (perennial)
Niche breadth	Broad	Narrow
Reproductive allocation	High	Low
Genetic variability	Low	High
Defensive allocation	Low	High
Mortality	Density independent	Density dependent
Population and Community characteristics:		
Total organic material	Low	High
Plant architecture	Simple	Complex
Species diversity	Low	High
Symbiotic associations	Few	Many
Chemical diversity	Low	High
Food webs	Simple	Complex
Population Dynamics	Fluctuating	Constant
Energetics:		
Net production (yield) per unit biomass	High	Low
Standing crop of biomass	Low	High
Biomass supported per unit energy flow	Low	High
Food chains	Simple, short	Complex, long

Nutrient flux:		
Element cycles	Open (leaky)	Closed (conservative)
Pathways for chemical transformations	Simple	Complex
Exchange rates	Rapid between soil and plant biomass	Slow, efficient
Control	Abiotic, anthropogenic	Biological

An important aspect that cannot be overlooked, relates to the third trophic levels, because in agroecosystems predators and parasite guilds are frequently depressed by cultural practices. This naturally distinguishes the impact of higher trophic levels in controlling pests. However generalist predators such as lady bird beetles, ground beetles, lacewings, anthocorids, reduviids and spiders can play an important role in keeping down pest population if they are provided with suitable habitats and an insecticide free environment.

Growing two or more crops together or multicropping presents a more useful habitat for insect pests, predators and parasites, with increase in the number of their interactions. In addition it has also been found that significantly higher densities and species richness of macroarthropod predators inhabiting the soil, particularly ground beetles and spiders occurred in intercropped areas and supported more complex systems. Multicropping with legumes helps restore more natural ecological inputs to agroecosystems. Interplanting maize with soyabean and groundnut, wheat and soyabean, maize and green gram are some of the methods in vogue.

Another aspect pertains to the introduction of biological control agents which in many cases have resulted in the extinction of native species in Hawaii. Bennet (1993) discussed that a more recent introduction in Mexico, of the parasitoids *Cotesia flavipes* on the sugarcane bore *Diatraea saccharalis* has apparently displaced the native parasitoid *Apanteles diatraeae*. Similarly the ability of an introduced biological control agent to disperse beyond the initial target area must be given active consideration. For instance, *Cactoblastis cactorum,* the caterpillar used to control cactus *Opuntia* sp. spread from the Carribean where it was established for biological control, to Florida where it threatens the survival of a rare cactus. The impact of a high order predator has resulted in reducing the efficiency of predator like *Chrysoperla carnea* the neuropteran, as a biocontrol agent for the cotton aphid *Aphis gossypii*. All the same it cannot be denied that the permanence of biological control is effectively irreversible and it would continue to be effective for long. In short "biological control is an arrow in the quiver of pest management, perhaps the only arrow in some cases of pests of environmental concern (Strong, 1997) (Fig. 9.9).

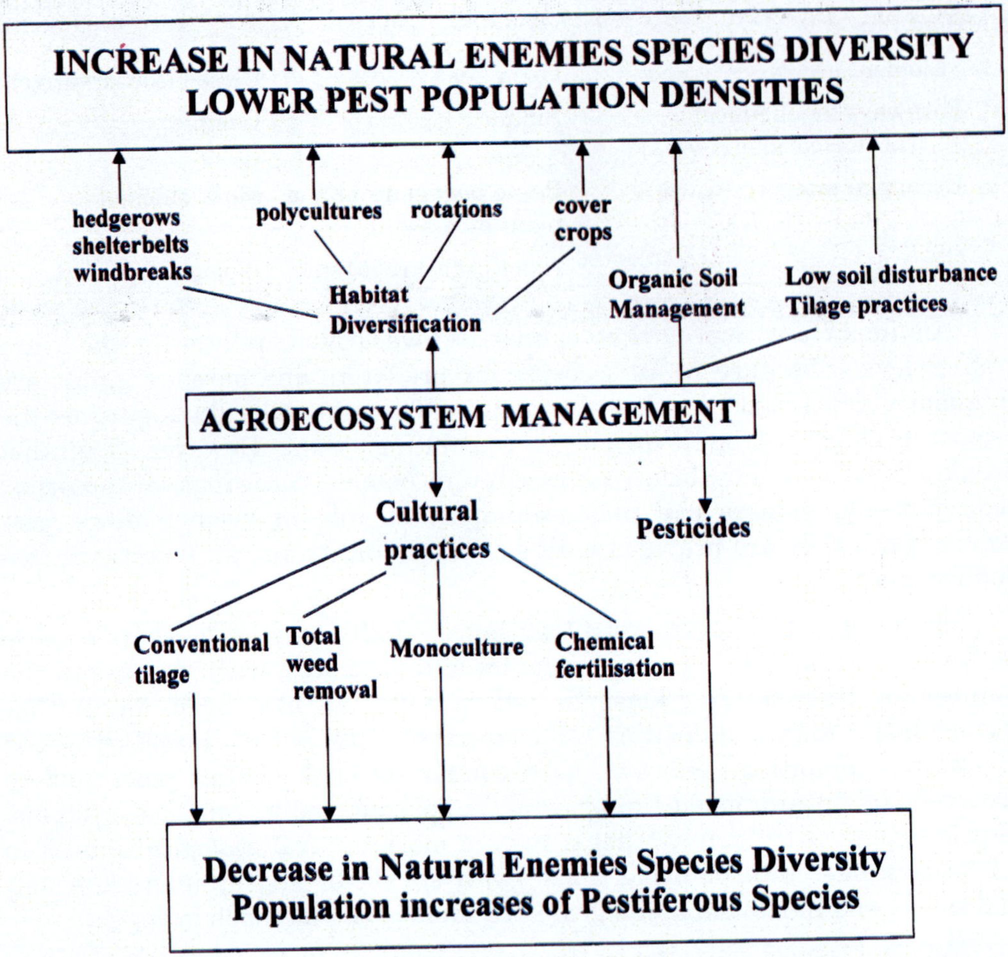

Figure 9.9. Biodiversity linkages in Agroecosystems (After Altieri and Nicholls, 2002)

Sustained biodiversity in agriculture

The use of methods to increase agroecosystem biodiversity through good management practices for enhanced biodiversity is called for, which can lead to sustainability of agroecosystems through providing adequate ecological services such as biological control, nutrient cycling and water and soil conservation (Fig. 9.9). Regulating insect pests through healthy soils and rich natural enemy biodiversity are important aspects. In monoculture crops such as rice, wheat, sugarcane there is a necessity to reintroduce biodiversity by introducing marginal vegetation which can serve as biological corridors allowing the movement and distribution of biodiversity. These field margins are reservoirs of natural enemies of the crop pests. Whenever crops are bordered by vegetation, increased abundance of natural enemies are evident. They may also serve as overwintering sites for natural enemies and also provide increased reservoirs such as of alternate prey, pollen and nectar for parasitoids and predators (Altieri and Nichols, 2002). The development of molecular techniques to manipulate genomes of crop plants and natural enemies are among the new insights into biocontrol practices. While transgenic plants may increase biological control efficiency, the development of transgenic natural enemies to tolerate pesticides and adverse climate are also important.

Tropical agriculture is characterized by employing more than one crop in sequence, combination or both and crops with annuals as well. In Africa 98 % of all cowpeas, a vital legume, are grown in combination with other crops. Farmers often mix strains of a given crop in their fields as a hedge against the vagarious of weather. They also tend to recognize the dependence of their farms on adjacent ecological systems and to tolerate wild plants on the outskirts of their fields. It is believed that crop relatives whose continued interbreeding with domestic descendents contributes to genetic variety. More commonly farmers mimic the species diversity of natural systems and grow varied crops together; such multiple cropping still accounts for as much as 20% of the world food production.

Biodiversity linkages in Agroecosystems

(e.g. Rice Field Ecosystem)

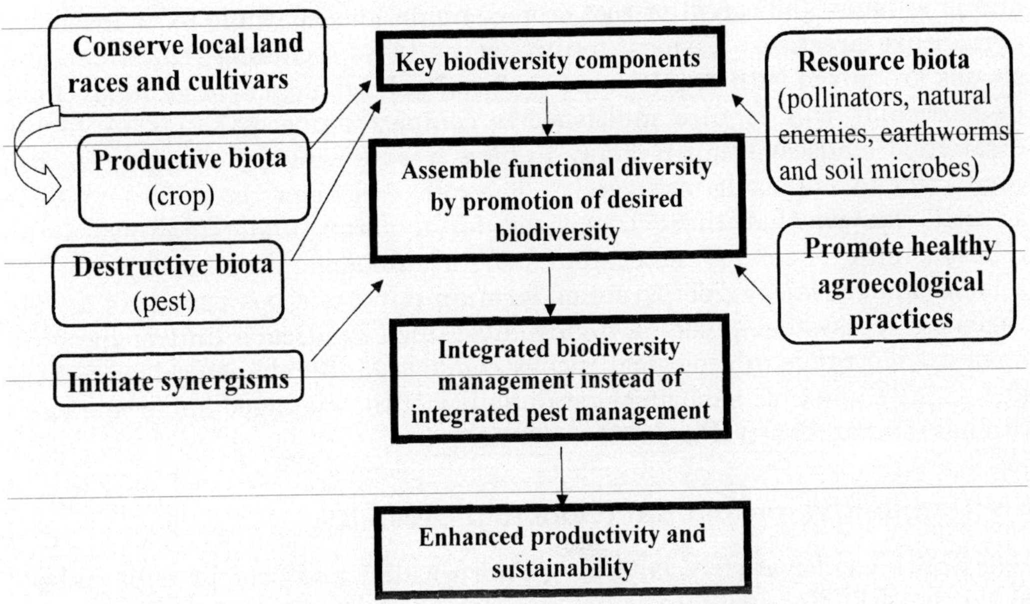

Figure 9.10. Agroecosystem Management

Some of the cropping patterns are - *Multiple cropping*: Growing two or more crops on the same field in a year; *Intercropping*: Growing two or more crops simultaneously on the same field per year. There is intercrop competition during all or part of crop growth and farmers manage more than one crop at a time in the same field. *Mixed intercropping*: Growing two or more crops simultaneously with no distinct row arrangement. *Row intercropping*: Growing two or more crops simultaneously with one or more crops planted in rows. *Strip intercropping*: Growing two or more crops simultaneously in different strips wide enough to permit independent cultivation but narrow enough for the crops to interact agronomically. *Relay intercropping*: growing two or more crops simultaneously during part of each one's life cycle. A second crop is planted after the first crop has reached its reproductive stage of growth but before it is ready for harvest. *Sequential cropping*: growing two are more crops in sequence in the same field per year. The succeeding crop is planted after the preceding crop has been harvested. *Double cropping* is growing two crops a year in sequence.

The cultivation of multiple species by farmers involves biodiversity issues, since cropping combination represent a planned biodiversity strategy which influence increased productivity. Increased use of fertilizers and land use practices are biodiversity-linked, since it is useful in sustaining biodiversity in agriculture. The planting of animal and perennial crops between rows of nitrogen fixing trees and shrubs meant to check soil erosion, also promotes biodiversity. Intercropping and crop rotation practices with combination of cereals and legumes, sorghum with pigeon pea, pearl millet and groundnut and cereal or root crop combinations continue to be intensive agriculture practices. Besides, multipurpose trees including fruit trees and shrubs are mixed with arable crops on farmland. Management of biodiversity in particular will require multispecies complex agroecosystems. A highly diversified landscape unit is likely to have a wide range of ecological niches conductive to enhancing biodiversity. The role of biodiversity in agriculture is so little known that there is a need for a better understanding of the relationship between agroecosystem structure (biodiversity) and function. In a landscape mosaic, agroecosystems form an important component of a wide variety of ecosystem types. A highly diversified landscape unit is likely to have a wide range of ecological niches conductive to enhancing biodiversity and at the same time ensuring sustainability of the managed landscape itself (Ramakrishnan, 2002) (Fig. 9.10).

Shifting Cultivaton and Entomofaunal Diversity

Shifting cultivation, a land use pattern widely practiced by tribal people of Northeast India causes large-scale damage such as reduction in diversity of vegetation, degradation of soil fertility and consequently decline in productivity of trophical rainforests. Shifting cultivation practice like repeated burning and cropping leads to a decline in rebuilding of nutrients and organic matter in soil. The entomofauna in natural forest litter ecosystem involving diverse arthropods have significant role in increasing the humus and soil fertility by way of leaf litter breakdown. Rapid shortening of follow period (time between two successive cropping) tends to be a major problem in maintenance of soil fertility in slash and burn agriculture. The degree of diversity of litter entomofaunal populations depends on the nature of the substrate, with changes on the relative consistency of specific composition of species. The role of soil fauna in litter decomposition is well known and some studies like differences in arthropod population structure in soils of forests and shifting cultivation area and changes in termite species composition due to shifting cultivation provided, information on the quantitative and qualitative abundance of litter insects including prey-predator resources. The litter habitat provides a large and varied assortment of ecological niches enabling different arthropod groups to occupy a variety of tropic positions. In a forest ecosystem, the litter inhabiting arthropods play a major role in the breakdown of leaf litter which in turn increase the humus and soil fertility (Seasted, 1984). Comparision of entomofauna associated with litter of various fallow sites showed considerable variation both in terms of species diversity and density. Beetles, collembolans, ants, mites and spiders are particularly important in forest litter decomposition (Kevan, 1962). Return of subsoil material to the forest floor in small litter concealed mounds by ants is an important process. Saprophagous mites constitute one of the most important group of Archnidae which helps in decaying through their feeding activity on leaves, wood and fungal hyphae (Wallwork, 1970).

CHAPTER -10

THE TROPICAL FOREST SCENARIO

Nature reaches its fullest expression in the tropical forest both in the number of species and complexity of their interactions. Tropical life has evolved an exuberant variety of species with the diversity of trees, birds, insects and other which has resulted in the formation of intricate food webs and other intraspecific interaction between two or more species including competition, mimicry, parasitism, mutualism. The diversity of niches or the range of physical and biological conditions permitting a species existence is considerable. Structurally complex vegetation might accommodate large numbers of bird niches and with a wide range of food resources available, birds have varied opportunities to specialize in their feeding habits. Habitats that offer a wide variety of foraging strategies would be able to accommodate the feeding strategies of a large number of birds. By offering a greater number of microhabitats such as the canopy, understorey, epiphytes and herbaceous vegetation on the ground, the forests frequently tend to support more species. Species guilds are very typical of tropical forests, guilds being species that share a common livelihood. For instance, guilds may be united by their diet, such as those community feeding on nectar. In other words they are structurally diverse, but functionally similar. The effect is to isolate groups of species that resemble each other ecologically more than they resemble the members of other guilds.

The value of the guild concept is that it allows comparison of communities composed of species that are taxonomically unrelated to one another. Evergreen tropical forests produce many kinds of food resources that flourish all year around such as fruits, nectar, seeds, besides animal resources. Members of the tropical ommnivorus guilds are versatile, offering opportunities for dietary specialization such as for cracking seeds, sucking nectar or consuming insects. Undoubtly the tropical forests offer large guild niches and the tropical guilds are more tightly packed with species. Many factors contribute to the greater diversity of the tropical forests. These include a more complex structural habitat, additional guilds occurring in changing environments such as for arboreal seed eaters, terrestrial frugivores, great plant diversity providing resources for humming birds feeding on nectar and frugivores, besides close packing of many guilds which reflect exhuberent speciation.

Figure 10.1. A Natural Forest

The *forest canopy* harbours the largest diversity and being called the last biotic frontier or power houses of plants, the canopy receives the maximum sunlight and hence with increased photosynthetic capacity it harbours diverse guilds. The canopy often reaches over 50 metres with broad gaps through which sunlight reaches the foliage below. The dense foliage commonly lies at heights of 20-30 metres and many species of trees coexist at this level (Fig. 10.1). Below this second tier, other small plant takes their place in vertical sequence – small trees, shrubs and herbs. To support the increasing cost of respiration as a result of expanding superstructure, a plant must continuously increase the amount of foliage, and most of the plants do this by expanding the size of their crowns. The canopy and understorey present very different environments. The huge crown and the species diversity of the canopy have enabled exploitation of the canopy by pollinators, frugivores and leaf feeding insects and being a patchy environment, they must move over longer distances. Structural differences between the canopy and understorey relate to microclimatic differences that affect animals or plants. The top of the canopy is hot, relatively dry and windy by day. In contrast, the environment near the ground is perpetually humid, poorly illuminated and practically windless. Because of these differences, the canopies and understorey are regarded as different domains by many of the forest animals. Numerous mammals, birds, reptiles and arthropods spend their entire lives within one vertical zone, other zones being exploited by different species. The diversity of species is higher, nearly twice as high with lower and middle portions of the canopy than below in the understorey. The depth of the species foraging zones is correlated with the height of the forests. Species living near the ground or in the topmost canopy have less latitude to vary their foraging heights, than those living midway between the two extreme zones. Accordingly the species occupying the middle tiers of the forest exhibit broad foraging zones compared to those living near the ground. Many factors continue to contribute to the extraordinary diversity of the tropics including a more complex structural habitat, greater tropical plant diversity and the close packing of many guilds. Such finely tuned ecological roles are very typical of species living in diverse tropical communities. A number of additional mechanisms may also contribute to tropical tree diversity, and these include a great variety of seed predators and dispersers, lighter species packing, a larger regeneration guilds and more complex microsuccession (Terbrough, 1992). Further, the vertical dimension of tropical forests contains a complex gradient of light conditions so that species are adapted to occupy specific positions. Tropical forests are able to achieve an impressive status by efficient uptake and retention of mineral elements, with many plants having evolved a whole series of adaptations to concentrate and conserve nutrients. In short, tropical forests are the most productive ecosystems and carry more leaves per unit area than other forests. Productivity is concentrated in the upper canopy where the crowns are fully exposed to sunlight. The fact cannot be derived that monoculture crops have restricted the depth of biodiversity in forests. The ravages caused by the teak defoliater *Hyblaea puera* is well known (Fig. 10.2)

Forest Canopy Diversity

The upper tree canopy of many forest ecosystems harbour diverse communities, particularly of vascular epiphytes and arthropods. Epiphytes contribute not only constitute to the diversity of birds, but also function as a 'keystone resource', and such genera as *Rhododendron*, *Anthurium*, *Peperomyia* are known. The many microhabitats and substrate types existing within the canopy contribute to biodiversity, with ants as the most dominant group, followed by beetles, oligochaetes, isopods, mites and collembola. At the community level, canopy organisms are of interest, since they are members of many symbiotic associations and provide a source of food and shelter for canopy animals. In many tropical forests the epiphyte community provides energy, nutrients, water and nesting materials, in particular for birds which include humming birds, fly catchers, finches and the like. There is a higher diversity in the tropical forest structure, besides the presence of certain resource elements. Epiphytes therefore augment bird species diversity by swelling the canopy resource pool, enhancing greater opportunities for greater bird diversity. Besides, certain epiphytes such as mosses, bromelids and canopy humus provide microhabitats for invertebrates. The carrying capacity of bird and mammalian frugivores in tropical wet forests may depend upon certain keystone plant resources that provide food and energy resources for maintaining animal populations. Epiphytes are keystone species because they relatively function in recruiting nutrients from atmospheric resources.

Figure 10.2. Teak trees defoliated by *Hyblaea purea* (inset)

Hotspots are not only vital areas of species-level endemism, but are also highly significant reservoirs of unique and threatened evolutionary history. Not only are the hotspots critical for a large portion of the diversity of primates and carnivores, but the evolutionary history of extinction-prone that species are disproportionately clustered in these areas. The majority of the species which are more likely to become extinct are endemic to the hotspots: viz. 77% of threatened primates and 60% of threatened carnivores.

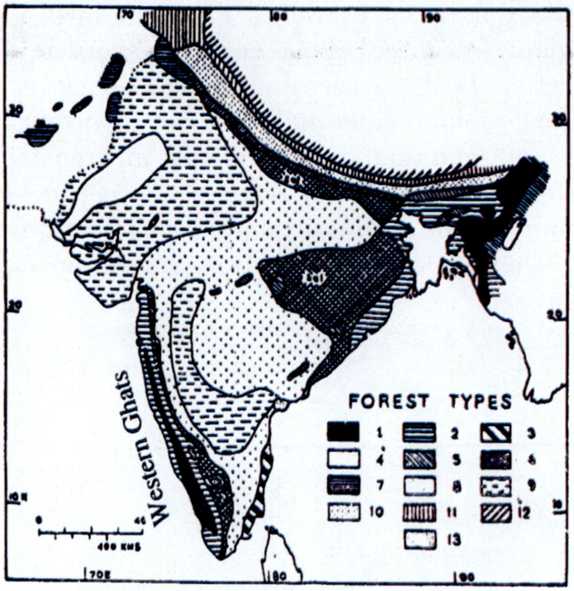

Figure 10.3. Western Ghats and Forest types

1. Tropical wet ever green; 2. Tropical semi ever green; 3. Dry ever green; 4. Tropical moist deciduous; 5. Sub-tropical wet deciduous; 6. Sub-tropical Pine; 7. Sub-tropical dry; 8. Dry deciduous; 9. Tropical thorn; 10. Moist temperate; 11. Dry temperate; 12. Alpine; 13. Tidal

The Western Ghats (Fig. 10.3)

The Western ghats and Eastern Himalayas are regarded not only as biological treasure groves, but are the also considered as the world's 25 biodiversity hotspots, since they are biodiversity rich areas which are under threat. Species diversity is unevenly distributed across the globe, with terrestrial diversity concentrated in a few restricted biodiversity hotspots. Conservation of these hotspots are critical because they harbour growing number of threatened species. In view of these mountain regions retaining much larger areas under natural or seminatural biological communities, both the Western Ghats and Eastern Himalayas are the most significant repositories of India's biodiversity. The levels of biodiversity vary along many gradients and diversity of plant species also increases as one crosses the ghats from the east to west. Another gradient relates to the increase in the number of plant species with an increase in the mean temperature from higher elevation to coastal plains (Gadgil, 1996). The diversity of trees is fundamental to total rainforest biodiversity, because trees provide resource and habitat structure for almost all the rainforest species. For example, the Agasthyamalai range located at the southern end of the Western ghats is known for high plant biodiversity habouring 2000 flowering plants. Tropical forests therefore represent are arena for many biotic interactions, well evident at a community level as plant-pollinator and plant disperser interactions. Among vertebrates 180 species of reptiles are known from the Western Ghats, of which nearly 50% are endemic, making the Western Ghats the richest biogeographic zone for reptiles in India. Of particular importance are the nearly all the 34 species of uropeltid snakes, 7 species of pit vipers and seven agamid lizard species. A large concentration of biodiversity rich sacred groves, besides the vegetation types of the Western Ghats, in particular the dominant flora and their distributors are indicted in Table 1.

Table 1. Vegetation types of the Western Ghats

Vegetation type	Distribution	Dominant flora
Tropical evergreen forests	200-1500 m ASL; 2500-5000 mm rainfall	Emergents upto 60 m; *Acrocarpus, Agalia, Artocarpus, Calophyllum, Canarium, Cullenia, Dipterocarpus, Holigarna, Knema, Myristica,* etc.
Moist Deciduous forests	500-900 m ASL; 2500- 3500 mm rainfall	*Bridelia, Pterocarpus, Sterculia, Pterospermum, Lagerstroemia, Tectona, Terminalia,* etc.
Dry Deciduous forests	300-900 m ASL; 1000- 2000 mm rainfall	*Albizia, Anogeissus, Bauhinia, Buchnania, Butea, Dillenia, Emblica,* etc.
Scrub jungles	200-500 m ASL; 300-600 mm rainfall	*Acacia, Carissa, Capparis, Flacourtia, Gardenia,* etc.
Sholas	Above 1500 m ASL; medium to high rainfall	Short trees, 15-20 m high; *Actinodaphne, Elaeocarpus, Eunymus, Michelia, Rhodomyrtus, Schefflera, Symplocos,* etc.
Savannas	1700-1900 m ASL; medium to high rainfall	Grass; *Chrysopogon, Arundinella, Eulalia, Heteropogon,* etc.
High rainfall savannas	Montane; Extremely high rainfall	Herbaceous to shrubby cover; *Ligustrum, Rhododendron, Anaphalis, Strobilanthus,* etc.
Peat bogs	Above 2000 m ASL; High rainfall	Grasses, sedges and mosses; *Carex, Cyanotis, Cyperus, Eriocaulon,* etc.
Myristica swamps	Sea level to around 600 m ASL, medium to high rainfall	*Myristica, Knema, Hydnocarpus, Lophopetalum,* etc.

(Source: Nair and Daniel, 1986).

Utkarsh *et al.* (1998) identified seven different types of vegetation that are distributed in species composition in the Western Ghats. These are, closed canopy evergreen, semienclosed canopy evergreen, stunted evergreen, semi-evergreen, moist deciduous, dry deciduous and scrub/savannah vegetation. Wild relatives of cultivated plants are concentrated in Western Ghats and their tract is a hotspot area with high biological diversity which is under constant threat. Studies on butterfly assemblages in the Western Ghats, have shown that the butterfly densities were highest in the deciduous forests. Natural vegetation types also show high beta diversity, but moderate to low species ubiquity. Although these natural types harbour many localized and some widespread species, generalist species invade along paths and canopy openings. The forests exhibit low to moderate levels of cohesiveness. The distinctness of the species composition of the semievergreen forests communities appeared low, next only to scrub/savanna. This is probably because these forests may lack their own distinctive species and by the very nature of their floristic characters draw the species from deciduous and evergreen forests as well as attract generalists. The high alpha and beta diversity in the semievergreen forests is to be expected in view of the possibilities of shared, widely distributed species.

Some of the common butterflies include species of the genera *Papilio, Troides, Graphium* (Papilionidae); *Catopsilia, Eurema, Prioneris* (Pieridae); *Melanitis, Athyma, Danaus, Junomia, Cynthia, Euploea* (Nymphalidae) *Leptotes, Chilades, Neopithecops, Absira* (Lycaenidae) and *Badamia, Notocrypta* (Hesperiidae). India however has a butterfly fauna of 1500 species, of which the Western Ghats harbour 330 species including 37 endemic species (Kunte *et al.*, 1999).

Animal biodiversity is very significant, in particular the occurrence of high levels of endemism in amphibians and reptiles. Nearly 2000 species of flowering plants and 285 species of vertebrates are endemic (Swengel, 1991). Among the vertebrate, endemism is highest in amphibians (87) species followed by reptiles (89), fish (84), birds (15) and mammals (12). Among the fishes, 56% are endemic and the genera of invertebrate are *Pristolepis, Badis, Psilorhynchus, Bharania, Travancorica*, besides *Puntius fasciatus, Etroplus canarensis, Silurus wynadensis*. 41% of the amphibian fauna are endemic, with the anurans (frogs and toads) represented by the genera *Bufo, Melanobatrachus, Microhyla, Ramanella, Uperodon, Micrixalus, Nannobatrachus, Nyctibatrachus, Rana, Rhacophorus, Polypedates* and the caecilians like *Icthyophis, Uraeotyphlus, Gegenophis, Indotyphlus*. On the whole, 117 species are on record under 21 genera. Mention has to be made of a unique frog *Nasikabatrachus sahyadrensis* which has led to the creation of a new family related to similar amphibians found as far many as Seychelles.

Among reptiles, of the 156 species on record, 93 are endemic, with 86 species of snakes, 63 of lizards, 6 of turtles/tortoises and 1 of crocodile. There are more than 500 species of birds and 12 species of endemic mammals. The insect and other invertebrate fauna are too diverse and more than 300 species of butterflies are on record, beside other insects.

Silent valley in the Western Ghats is a typical example of a tropical rain forest, consisting many of tropical evergreen, grasslands and shola vegetation with tremendous floral and faunal diversity. In the words of Salim Ali the great ornithologist, "Silent valley is not just an evergreen forest, it is a very fine example of one of the richest, most threatened and least studied habitats on earth" It harbours a large number of rare, endangered and endemic flora and fauna. These forests have remained as ecological islands preserving the relic characters of the fauna and flora. In the biogeographical sense, the silent valley represents an ecological island in the humid tropical forests of peninsular India. These forests are typical wet evergreen with microclimatic variations one of the hilly and undulating terrain with changes in species associating compositions. The major vegetation types are the moist deciduous forests, riparian evergreen forests, sholas and grasslands. Besides 100 species of butterflies and 400 species of moths, there are around 220 species of other insects. Thirty five species of reptiles, 19 species of amphibians, 192 species of birds and 35 species of mammals have been recorded, with many more available.

Figure 10.4. Lion-tailed macacque

Figure 10.5. Common langur

Among the mammals are there primates, the Bonnet macaque (*Macaca radiata*), the lion-tailed macaque (*Macaca sileneus*) (Fig. 10.4) and the nilgiri langur (*Presbytis johni*). The common langur *Presbytis entellus* (Fig. 10.5) is also evident. Among the large carnivores are the leopard (*Panthera pardus*), the tiger (*Panthera tigris*), the wild dog (*Cuon alpinus*), the jackal (*Canis aureus*) and the sloth bear (*Melursus ursinus*), the jungle cat (*Felis chaus*), leopard cat (*Felis bengalensis*), small Indian civet (*Viverricula indica*), the brown palm civet (*Paradoxyurus jerdoni*), the common mongoose (*Herpestes edwardsi*), the Malabar giant squirrel (*Ratufa indica*), the porcupine (*Hystrix indica*). The bats are represented by *Rhinolophus rouxi, Cynopterus brachyotis*. The endangered nilgiri tahr (*Hemitragus hylocruis*) (Fig.10.6), the spotted deer and black buck (Fig. 10.7), the sambar deer (*Cervus unicolor*), the barking deer (*Muntiacus muntjak*), the gaur (*Bos gaurus*) (Fig. 10.8), the Indian pangolian (*Manis crassicaudata*) (Fig. 10.9), the wild boar (*Sus scrofa*) and the elephant (*Elephas maximus*) (Fig. 10.10) are equally common.

Figure 10.6. Nilgiri Tahr

Figure 10.7. Spotted deer and black buck

Figure 10.8. Gaur

Figure 10.9. Pangolin

Figure 10.10. An elephant group

Of the several species of birds are the reptiles, pigeons, parakeets, cuckoos, wood peckers, drongo, mynahs, bulbuls, babblers, flycatchers, warblers and sun birds. The rose ringed parakeet (*Psittacula krameri manillensis*), the grey drongo (*Dicrurus paradiseus paradiseus*), the hill myna (*Gracula religiosa indica*), the black bulbul (*Hypsipetes madagascarensis*) the red vented bulbul (*Pycnonotus cafer*), the fly catchers (*Musciscapa nigrorufa*), the small sunbird (*Nectrinia minima*) are common. Besides these, the wagtail (*Motacilla indica*), common kingfishers (*Alcedo attahis taprobana*), the frog mouth (*Batrachostomus monilger*). Several raptors, parakeets, bee eaters, hill mynas occupy the upper side of the canopy. Bulbuls, warblers, drongos, barbets, wood peckers, flycatchers and sunbirds were frequent in the lower side of the canopy and middle storeys (Fig. 10.11). Babblers, thrushes, fowls were typical ground storey birds. On the basis of their performance for food, there are broad guilds like nectarivores, frugivores, omminivores and carnivores. Birds of prey such as eagles, hawks, falcons, kites are equally common in natural habitats, with species such as *Milvus migrans govinda*, the pariah kite, *Haliastur Indus*, the brahminy kite, *Falco peregrinus*, the falcon, *Hieraetus pennatus*, the hawk eagle, to mention some of the more common species out of around 20 birds of prey. Quails and jungle fowls are equally common, with such species as *Coturnix chinensis chinensis*, the blue breasted quail, and the gray jungle fowl, *Gallus sonnertii*.

Figure 10.11. Birds and Niche diversity
(From Birds of the Rishi valley by Rangaswami and Sridhar)

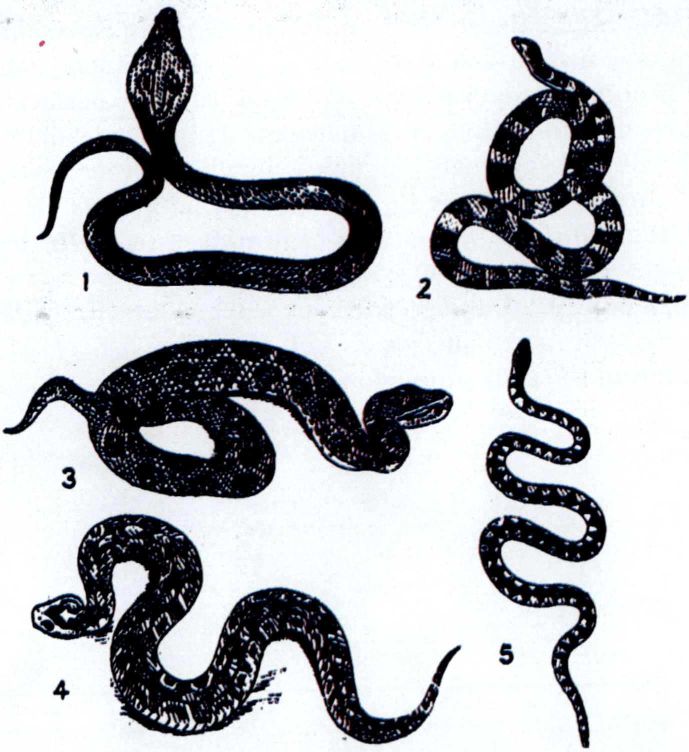

Figure 10.12. Common snakes
1. Cobra; 2. Krait; 3. Russel's viper; 4. The litteber Indian viper; 5. Sea-snake.

Some of the common snakes of the silent valley are the rat snake (*Coluber* (*Ptyas*) *mucosus*) and microhabitats like buttresses of large trees, thick layers of leaf litter, rock cervices and bamboo clumps offer good ranges for snakes such as pit vipers (*Trimeresurus malabaricus*) and *Hypnale hypnale*. A thick layer of leaf litter is ideal habitat for small terrestrial burrowing snakes like *Oligodon* and the uropeltids *Melanophidium punnctatus*, and keel backs *Macropisthodon plumbator*. Tree snakes include *Briga ceylonensis* and *Dendralaphis tristsi*. The spectacled cobra (*Naja naja*) and the king cobra *Ophiophagus hannah* and the krait *Bungarus caeruleus* and the viper, *Vipera russeli* are common (Manoharan *et al.*, 1999) (Fig. 10.12).

CHAPTER - 11

MARINE BIODIVERSITY

Introduction

India has a coastline of about 8000 km and a very wide range of coastal ecosystem such as estuaries, lagoons, mangroves, backwaters, salt marshes, sandy beeches and coral reefs which harbour unique faunal elements. The most extensive of all aquatic communities is the marine ecosystem involving the oceans. Life in the sea influenced by thermal variations and the geographical variations in the distribution of marine animals is the result of temperature variations. Pressure at various depths in the ocean induces a high degree of variability in animal life, besides adaptation against high salinity, and tidal influence, in particular at the sea shore where the low tide exposes the shore and high tide floods the substratum. The lives of shore inhabitants are distributed by the waves and they show suitable adaptations to withstand the same. An interesting phenomenon of the sea is upwelling, with the movement of water currents in a vertical direction, helping to bring nutrients to the upper surfaces of the ocean. Upwelling is important in organic production as the upwelled water bring nutrients of high concentrations. Upwelling is a major factor responsible for high plankton production. The basic factors influencing marine life therefore are light, temperature, pressure, salinity, tides and currents. Diurnal migrations of marine organisms are related to light and photoperiod and the colouration of animals depends on the availability of light. While many deep sea animals are colourless, some are dark brown to black.

Figure 11.1. Zonation of the marine ecosystem

Zonation in Marine Ecosystems

The biota of the ocean, as in freshwater ecosystems are classified on the basis of their habitats. The vast area occupied by the oceans naturally makes the zonation more conspicuous. The pelagic and benthic forms comprise the two main types of marine organisms. Organisms exclusively restricted to the open sea, and living with the aid of floating or swimming devices are *pelagic* their habitat being *pelagial*. Organisms which are dependent on the substratum, adopting a crawling, burrowing or sessile mode of life form the *benthos* and their habitat is termed *benthal*. The further classification of these zones is on the basis of the degree of light penetration. Accordingly the benthal zone is divided into *littoral benthos* and *abyssal benthal zone*. Littoral zones extend to a depth of 200 metres from the shore and may vary according to the steepness and slope of the shore into the sea. A gradual slope often results in a wider littoral zone. The terms continental shelf is applied to that portion of the land extending from the low tide mark up to where it drops suddenly towards the sea. The littoral zone is further subdivided into a *eulittoral* and a *sublittoral* zone. The tentative subdivision is based on the abundance or sparseness of marine phytoplankton. The lightless part of the benethal zone falls into two zones, the archibenthic zone starting from where the continental shelf ends and the abyssal benthic zone forming the sea floor. The open sea or the pelagial zone is divided into the neritic and oceanic regions. The neritic zone is the portion of open sea extending up to the limit of continental shelf and the rest is the oceanic zone (Fig. 11.1).

The littoral zone or animal diversity in sea shores

The littoral zone or the intertidal zone denotes the area in between the low and high tides, presently high fluctuations of environmental conditions, with a high rate of variability and periodicity of the fauna. Frequent covering and uncovering of the shore brought forth by the alternate ingress and egress of tides, repeatedly drying and wetting the zone, introduce quite a few problems. Animals consequently have to be adapted to withstand desiccation. Littoral polychaets escape desiccation by resorting to a tubicolous mode of life whereby they can resist desiccation by retaining sufficient moisture till the advent of the next wetting. Barnacles can survive desiccation by remaining in a state of suspended activity during dry period. The opercula of molluscan shells cover the occupants saving them from any possible dry up.

Yet another hazard of the shore environment relates to the action of waves. With tremendous force they lash on the littoral habitat and organisms in order to survive must be adapted to withstand the shock from the smothering of waves. From this point of view the shingle beaches where there is a continuous crushing and rolling of stones one over another, harbour scanty fauna. But animals in rocky beaches have a covering of sea weeds which absorb the shock of the wave impact. The crevices of rocks act as a good place of escape for the shore animals from the fury of waves. Abundance of fauna in the land ward slope of a rocky coast is also due to this limiting factor. Many sedentary shore animals adopt special modes of adhesion which keep them firmly attached to the substratum. Sponges and tunicates, for example, remain permanently fixed and when detached from their substratum die. The often flattened bodies of the planarians, and nemertines, as also the encrusting Polyzoa, sponges, etc. appear well suited to minimize wave action in many shore animals. But the situation is different in sandy beaches as there is a constant crowing and depositing action by the waves. Animals with firm attachments are misfits in this environment due to the loose nature of the substratum. Hence a majority of these animals have taken to a burrowing mode of life to escape wave action and fascinating modes of burrowing are met with in sandy beach dwellers.

The influence of varying salinity and temperature accompanying the low tides and high tides is also an important factor affecting the life of littoral animals. Surfaces of rocks, especially when they are bare, get heated up to a very high degree. Similarly sand gets heated up fast and cools down rapidly. Animals living between the sand particles are the *interstitial fauna* which comprise a group, individuals surviving or gliding in the interstiles of space between the sand grains. Three major types of littoral areas are recognized, the *rocky*, *sandy* and the *muddy* with animals occupying each type of shore exhibiting specific adaptations. Rocky shores are dominated by sessile and attached animals that resort to filter feeding. Some animals have developed an encrusting habit especially the bryozoans, coelenterates and some tunicates. They possess body structures which tend to reduce the wave action to a minimum. Those animals which grow within the crevices of rocks assume a bushy growth as in hydroids and bryozoans. Some others lie in calcareous tubes which they construct or by burrowing into the substratum. What is exposed outside is only their oral region which is provided with a filter-feeding apparatus.

Growing abundantly on the rocky coasts are the acorn barnacles which occupy the upper limit of the high tides. Further down, these barnacles are replaced by other species of barnacles. The presence of a solid substratum affords greater chances for attachment and therefore rocky shores have an abundance of sessile animals. Limpets like *Patella*, *Haliotes*, *Fissurella* and oysters, barnacles, tunicates and bryozoans which are mostly sedentary in their mode of life, inhabit the rocky substratum. Sessile organisms like the sponges, the colonial hydrozoans, anthozoans like *Gemmaria* and *Zoanthus* attached to rocks bear enormous colonies of these sedentary forms. Some rocky shore inhabitants have also taken to a boring mode of life. Unlike as in a soft substratum, rocks are hard and tough imposing a restriction on the number of such species to just a few. Soft rocks like limestones are generally preferred for boring by some sponges, annelids and molluscs, which are among the commonest of such forms, besides offering protection to the borers, these crevices act as places of escape for many forms after they have been deserted by the borers. Tubes of many tubicolous polychaets are often inhabited by other associated organisms (Fig. 11.2).

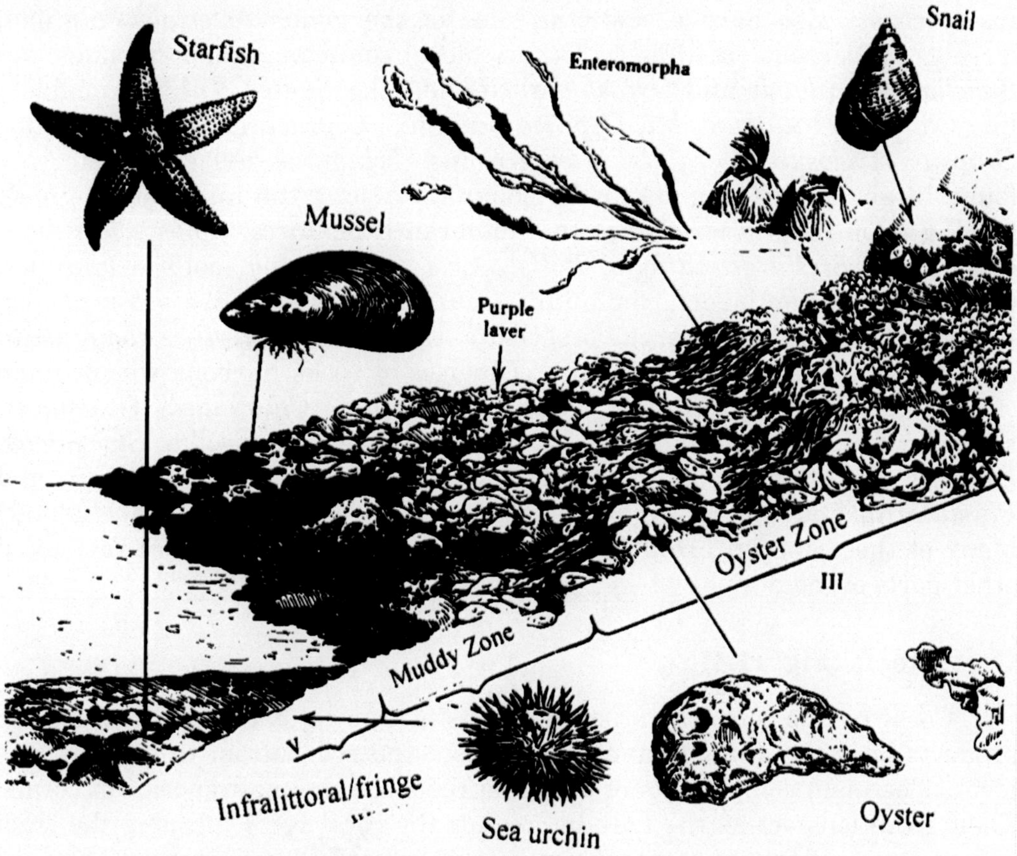

Figure 11.2. A rocky coast showing animal diversity

The zone of bivalves of which the first one is generally that of mussels like *Mytilus*. This zone is a vulnerable zone being exposed to the brunt of wave action and active predation by starfishes and gastropods. These bivalves grow in beds which act as a basis for tunicates, sponges and other small mussels. The next one, the zone of oysters, is a less vulnerable, zone while the third zone is that of the boring bivalves. They find protection from waves and predators by remaining within the crevices and exposing only the siphons. Certain other bivalves, especially those that are incapable of boring, live within the tubes built by others. Borers may resort to boring by chemical secretions or though mechanical devices. Mechanical borers are the commonest among thee and are chiefly represented by the genera like *Pholas, Martesia, Hiatella, Tridacna* etc. These boring animals are the prominent dwellers beside the sessile forms. Besides the bivalves a variety of polychaets live in tubes. *Serpulids, terebellids, sabellids* and *cirratulids* are the commonest among them. Bryozoans and brachiopods are other filter feeders found in the rocky intertidals. Many other animals graze upon the algal encrustations, the majority of these grazers being gastropods and chitons. Rocky coasts also have a few starfishes of the genus *Asterias, Palmipes,* largely predaceous in habit. Other more common fauna include are *Stoichactis, Ixalactis* and *Gyrostoma.* Nereids like *Eunice, Tubifex*; molluscs like *Chiton, Nerita rumphii, Cypraea ocellata, Cypraea arabica*; crabs like *Grapsus strigosus, Atergatis integrerrimus, Pilumnus vespertilo,* etc. are found in abundance (Fig. 11.3). Predominant among the numerous animals grazing upon the sea weeds are the nudibranch molluscs, *Aplysia benedicti, Onchidium, Elysia grandifolia, Eolis, Doris,* the floating mollusc *Ianthina,* etc. The midlittoral reef community consists of fishes like *Epinephelus, Petrachus,* the crabe *Pilumnus vespertilio,* the annelid *Eurythoe complanata* and molluscs like *Murex, Bursa* etc. Crevices of rocks harbour within them crabs like *Atergatis, Grapsus, Pilumnus, Matuta, Charybdea, Gelasimus*; squillids like *Sesarma* and *Gonodactylus*; coelenterates like *Stoichactis, Ixalactis, Zoanthus, Lobophytum*; the hydroids *Lytocarpus, Sertularia* and *Plumularia*; and annelids alike *Serpula, Polynoe, Eurythoe, Sabella,* etc. Many of these species are equally characteristic of the intertidal regions of other parts of India.

Coral Reefs (Fig. 11.4)

Coral reefs are characteristic features of tropical waters. These are the products of living animals and extend from the sea bottom to the low tide level. The main components of the coral reefs are the madreporarian corals. Their secretions act as the foundations for the coral reefs. Among the chief donors for coral formation three families are of prime importance. They are the Poritidae (*Porites minicoiensis, Porites mannarensis*), Acroporitidae (*Acropora intermedia*) and Astraeidae. Milliporidae also contribute substantially to coral reef formation. Secretions of calcareous algae go into the building up of reefs. Equally important are the calcareous Bryozoa and the calcareous masses such are further enlarged by the action of Foraminifera.

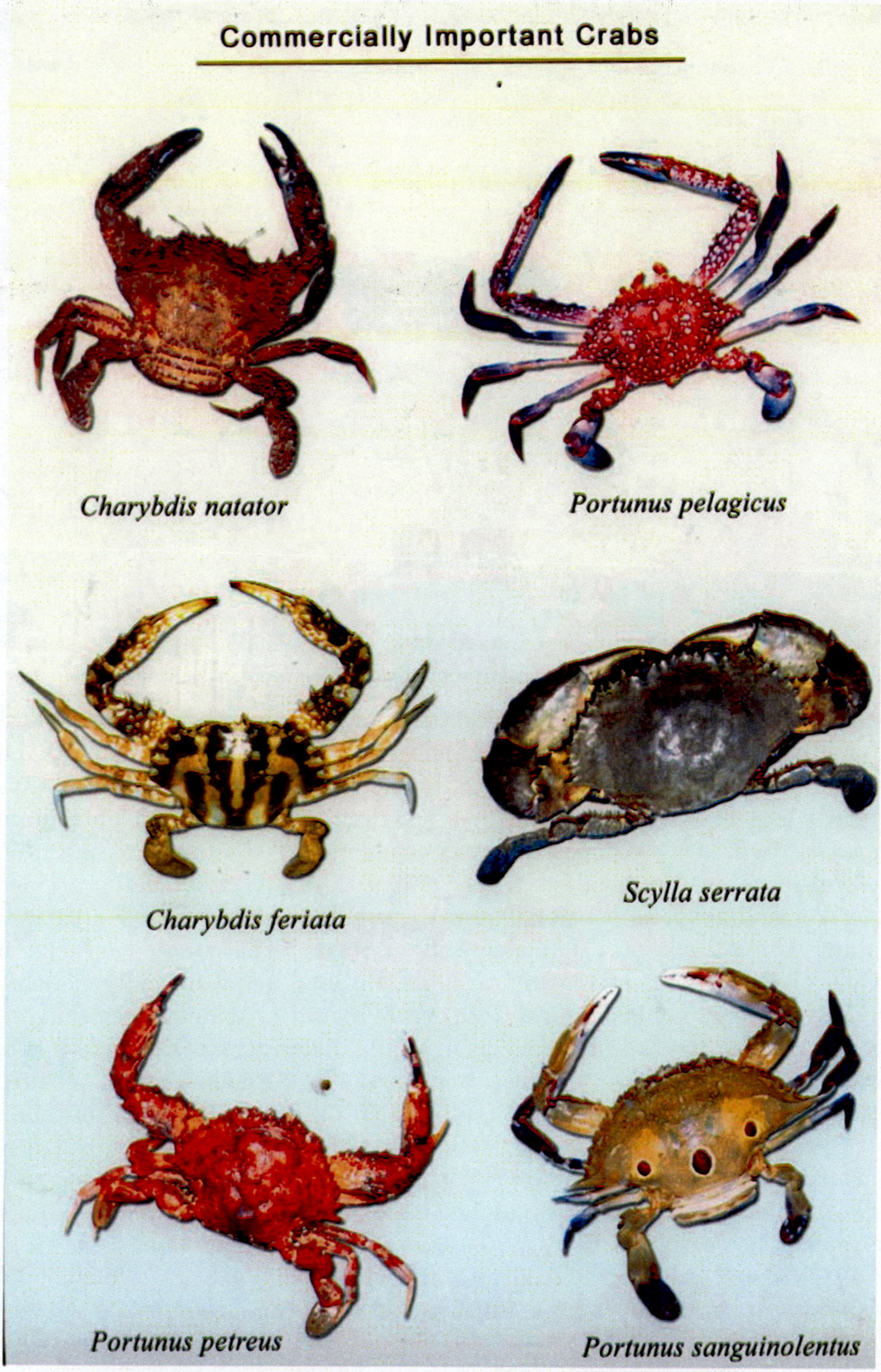

Figure 11.3. Common crabs (Courtesy, K. Venkatraman)

Figure 11.4. Coral reef

Three types of coral reefs are distinguished: fringing reefs, barrier reefs and atolls, fringing reefs, as in Mandapam, Appa Islands, and the Hare Islands outline the coasts and are only a few meters wide. Barrier reefs, as the great barrier reef in Australia, have a canal or lagoon between them and the coast. *Atolls* are ring shaped or horse-shoe shaped islands. The fauna of the coral reefs has no parallel for its number and diversity. Being a place of better concealment and safety, many animals thrive successfully, often escaping the attack of predators and enemies. The smaller cracks and crevices give shelter to retreating animals. Preying upon the indigenous coral fauna are a variety of predators. Coral reefs include a wide range of diversity with 32 of the 34 animal phyla present, in contrast to only 9 phyla represented as free living species in tropical rain forests. This is because of the high productivity of the coral reefs which produce 2500 gram of biomass per sq.meter per year, in comparison with 125 gm^2/year in the open ocean. The clarity of the water in the reef ecosystem with increased sunlight enables increased photosynthesis in the algae living mutualistically inside the coral. The world's largest coral reef is Australia's Great Barrier Reef with an area of 3,49,000 km^2 with over 350 species of corals, 1,500 species of fishes, 4,000 of molluscs, 5 species of turtles and provide breeding sites for 252 species of birds (IUCN/UNEP, 1988).

Pocillopora damicornis

Montipora aequituberculata

Favia pallida

Figure 11.5. Some common corals (Courtesy, K. Venkatraman)

Several islands in the Gulf of Mannar and nearabouts such as Krusadi group of islands, are coral islands with sand-stone forming the base rock. Side by side with an amazing array of corals a number of invertebrate types as well as brightly coloured coral fishes occur. Dominant corals of the areas are *Porites, Montipora, Poecilopora, Acropora, Madrepora, Favia, Symphylia, Psammocora, Platygyra, Merulina* and *Leptastrea*. Majority of these corals are found on the shore ward face of the rock or in the interior of the reef. Some are even found to occur in the rock pools towards the shore. *Porites* a massive coral seems to prefer the reef margin that is subject to the action of low tide waves (Figs. 11.5; 11.6). In the Krusadi area, a luxuriant growth of sea weeds is observed, the main species being *Sargassum, Caulerpa, Turbinaria, Spatoglossum, Gracilaria, Ulva, Enteromorpha, Chaetomorpha* and *Polysiphonia*. The dead coral rocks with plenty of openings harbour within them sponges, such as the massive sponge *Spirastrella incorstans* and *Callyspongia fibrosa*. Generally dominating sponges of the coral reef are found attached to the coral rocks or to other hard areas below and the sandy mud above. The live and dead corals afford shelter for a variety of sponges, coelenterates, planarians, polychaets, crustaceans, molluscs, tunicates and polyzoa. Coelenterates are generally represented by *Syncoryne*, *Sertularia*, *Eudendrium*, *Pennaria*, *Tubularia*, *Thyroscypus*, *Plumularia*, etc. which are reported to form encrusting hydroid colonies. The underside of stones harbours alcyonarians like *Clavularia*, *Alcyonium*, *Lobophytum*, *Sclerophytum*, *Telestor* and Zoantharians like *Zoanthus*, *Gemmaria*, etc.

Fungi scutaria **Euphyllia glabrescens**

Figure 11.6. Some common corals (Courtesy, K. Venkatraman)

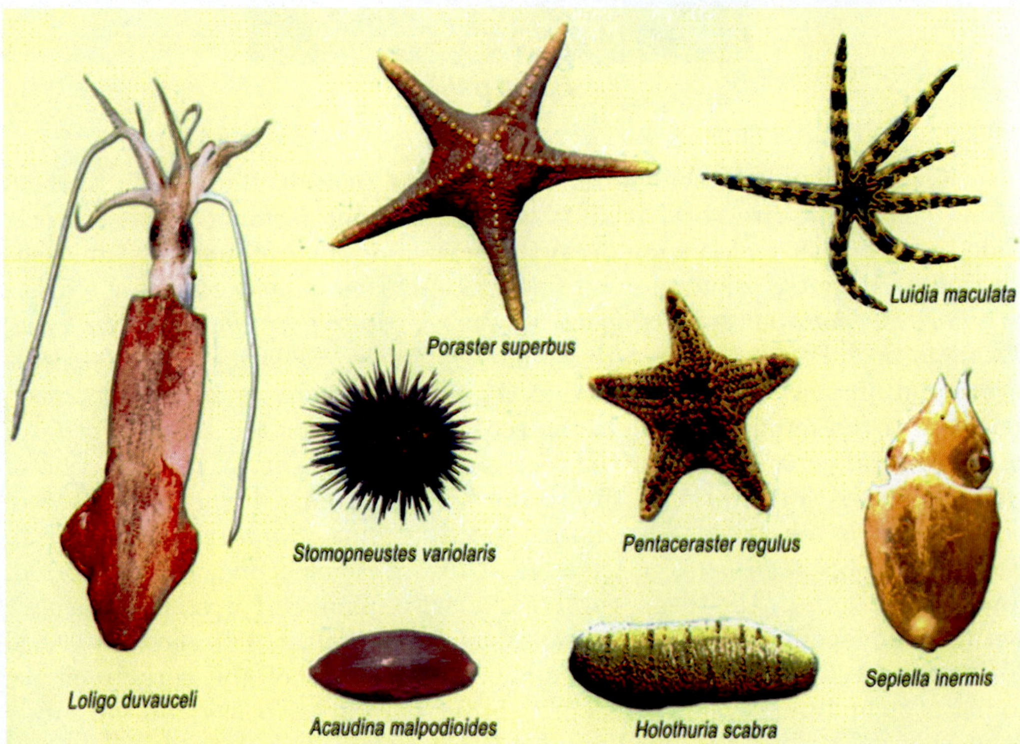

Figure 11.7. Common Echinoderms (Courtesy, K. Venkatraman)

The interstices of the coral reefs provide abundant and varied habitats for cryptic species, such as the countless number of polychaets, small crustaceans, bryozoans, hydroids and other smaller invertebrates. The annelid fauna is most abundantly represented by the polychaet families. *Nereis, Polynoe, Eunice, Lepidonotus, Iphione, Eurythoe* and *Hesione* among the annelids, and *Dendrostomum, Aspidosiphon* and *Phascolosoma* among the sipunculids are most commonly represented. Nudibranchs like *Doris* and *Dendradoris,* besides *Acanthochiton, Ischnochiton, Haliotes, Emarginula, Trochus, Turbo, Nerita Natica, Cypraea,* and *Thais* are the chief representatives of the phylum Mollusca. Under the coral rocks of the reef flat *Xancus pyrum,* the sacred conch shell is obtained. Among the echinoderms *Holothuria atra, H. scabra* and *H. edulis* are the chief occupants of the reef community. The echinoderms such as *Luidia, Pentaceraster, stomopneustes, Holothuria* are equally common (Fig. 11.7).

Beautifully coloured fishes like *Holocentrus, Lutianus* sp. *Pomacanthoides, Chaetodontopsis, Chaetodon, Linophora, Thallasoma* and *Zanclus* and the eels *Muraena tussellata, Gymnothorax undulata* and *G. punctatus* commonly frequent the coral reefs. The inhabitants of the reef community, when exposed at low tides are subjected to predation by a large number of birds like the crested tern *Sterna bergii,* the eastern golden plover *Pluvialis dominica fulva,* etc.

One comes across the maximum taxonomic diversity and biological productivity in the coral reef communities. Coral reefs are complex ecosystems with a large number of green plants contributing to its biomass. Plants and animals live in mutual union in a coral community. Tissues of the polyps of coral forming organisms contain within themselves the green algae namely Zooxanthellae. Zooxanthellae live in symbiotic, often mutualistic relationship with the corals. In return for the food they prepare from sunlight the algae receive the essential elements from the corals. Besides living as endosymbionts, a few species are recorded as ectosymbionts living as endosymbionts, a few species are recorded as ectosymbionts living over the limestone substrates. Symbiotic algae provide shallow water corals with more than enough carbon compounds (energy sources) for metabolism, growth and reproduction. They recycle host wastes efficiently, the waste elements containing nitrogen, phosphorous are needed as building blocks for the corals.

Sandy Shores

Constant movements of sand caused by wave action is typical of sandy shores, with diurnal, seasonal and annual rhythms of tidal action. The principal animal groups of interstitial waters are the Nematoda, Turbellaria, Annelida, Gastrotricha and Acarina. Other groups are represented in smaller numbers. Protozoa are sporadically distributed. Hydrozoa, Turbellaria,

Nemertinea, Rotifera, Archiannelida, Polychaeta, Ostracoda, Halicaridae and Nudibranchiata generally inhabit the top 20 cm of sand between the low and mid tide levels of the beach. Species of Gastrotricha, Kinorhyncha and Isopoda occur at mid tide levels preferably in deep layers of sand. Oligochaets are sporadic and are found at different levels. Harpactcoid copepods are seen at all depths, most of them being restricted to mid tide levels. Copepods when present show a distinct zonation. Tardigrades occupy the region between mid tide and high tide levels.

Slow moving forms constitute distinct communities, while mobile species appear in a variety of places. But the chief factors affecting their distribution seem to be food, temperature, moisture and the texture of the substrate. The spionid annelids *Prionaspis krusadiensis, Aricia* and *Lumbriconereis* are found here. The above forms possess the capacity to secrete mucus which assists in keeping a cover of sand grains over the body for purposes of protection. *Eunice tentaculata* is common below the low water mark. *Glycera lancadivae, Glycera alba, Nerita bonnieri, Pisionidens indica, Pisione cymplex* and the archiannelids *Saccocirrus minor* and *Protodrilus* also occur, though not in abundance.

Among the crustaceans *Emerita asiatica* is the most prominent form. *Albunea symnista* is found burrowing itself into the sand with its two antennae held together for purpose of breathing. Amphipods like *Harpinia* sp., the isopod *Sphaeroma walkeri*, mysid shrimps *Gastrosaccus spinifer*, hermit crabs *Clibanarius aretheusus* living within the shells of *Thais* and *Cerithium*, the eight-oared swimming crab *Matuta citor*, the spider crab *Philyra scabruesula, Ocypoda platytarsis* and *O. macrocera* are all very common. Lamellibranchs like *Donax cuneatus* and *Pamphia textile*, burrowing snails like *Sinum neritoideum*, common gastropods like *Oliva gibbosa* and *Terebra* spp., *Bullia vittata* are also found along the coast. Besides the sand dwelling eel *Ophichtys orientalis* is also met with. Siphonophores like *Physalia, Velella* and *Porpita,* medusae like *Aequoria* and *Chiropsalmus* also occur stranded on the beach.

The Pelagic Zone

The open ocean or the pelagic zone covering a vast area and large volume is the most important ecological division of the marine environment. Animals of the open ocean encounter stable and uniform environmental conditions due to the continuity of the water of the sea. This continuity of sea water causes the individuals of the open ocean to stay away temporarily or permanently from any solid objects and they are provided with different locomotor organs. Many forms however prefer to float freely in the vast expanse of the sea and these constitute the *plankton.* Majority of the planktonic forms are small, less than 3 mm in size and are designated the *microplankton*, while the larger members form the macroplankton. Planktonic organisms occurring at a distance less than 60 meters comprise the *nannoplankton*. From the point of view of nutrition, planktonic organisms fall under two categories: (a) the *phytoplankton* including plants and other autotrophs, and (b) *zooplankton* comprising the animal members or the heterotrophs. The phytoplankton constitutes the chief producers of oceanic environment. The members of the plankton living in the euphotic or lighted zone of the pelagic realm constitute the *epiplankton*, while those that are confined to the aphotic zone form the *bathyplankton*. Among the planktonic animals some continue to spend their entire life as plankton (*holoplankton*), while others spend part of their life history as sessile organisms (*meroplankton*).

In almost all the food chains of the oceans the first link is the phytoplankton and the link immediately following it is inevitably the zooplankton (being holozoic its mode of nutrition) which depends on the phytoplankton for its survival. Distribution and abundance of zooplankton are maintained by the distribution of phytoplankton. Diurnal vertical migrations are characteristic of the planktonic organisms. Planktonic forms migrate towards the surface during the day and return when darkness sets in or they may move towards the surface with the approach of darkness and return with the break of light. Although the phenomenon is often described as a phototropic behaviour of the plankton, the essential causes of the process betray simplicity and are more complicated and difficult to analyze.

Records of the plankton of the inshore waters of South India shows an abundance of the flagellate *Noctiluca miliaris*. Other planktonic organisms included larval forms of coelenterates like the *Planula, Ephyra, Semper's* larva and *Cerianthus* larva along with the medusae of *Bougainvillia, Obelia, Charybdea, Rhizostoma, Rhopalina* and the ctenophores *Pleurobrachia* and *Beroe* Among the annelids larval polychaets dominate, although *Tomopteris. Alciopa* and *Autolytus* are also not scare. Several species of the chaetognaths like *Sagitta* and *Spadella* are important constituents of the plankton. Heteropod and Pteropod molluscs like *Cariniaria, Pterotrachea, Clione, Creseis* and larvae of echinoderms such as bipinnaria ophiopluteus, echinopluteus, auricularia and the torniaria of *Balanoglossus* also dominate. Principal among the copepods are species of *Acartia, Calanopia, Labidocera, Centropages, Paracalanus, Canthocalanus, Acrocalanus, Eucalanus* and *Microsetella*. Although decapods are represented largely by larvae adults of Lucifer may also abound. The prochordate families like Larvacea, Thaliaceae and Ascidiaceae are common, with *Thalia democratica, Jassis zonaria* and *Salpa cylindrica* along with the cephalochordate *Amphioxus*, being often encountered in the plankton (Fig. 11.8).

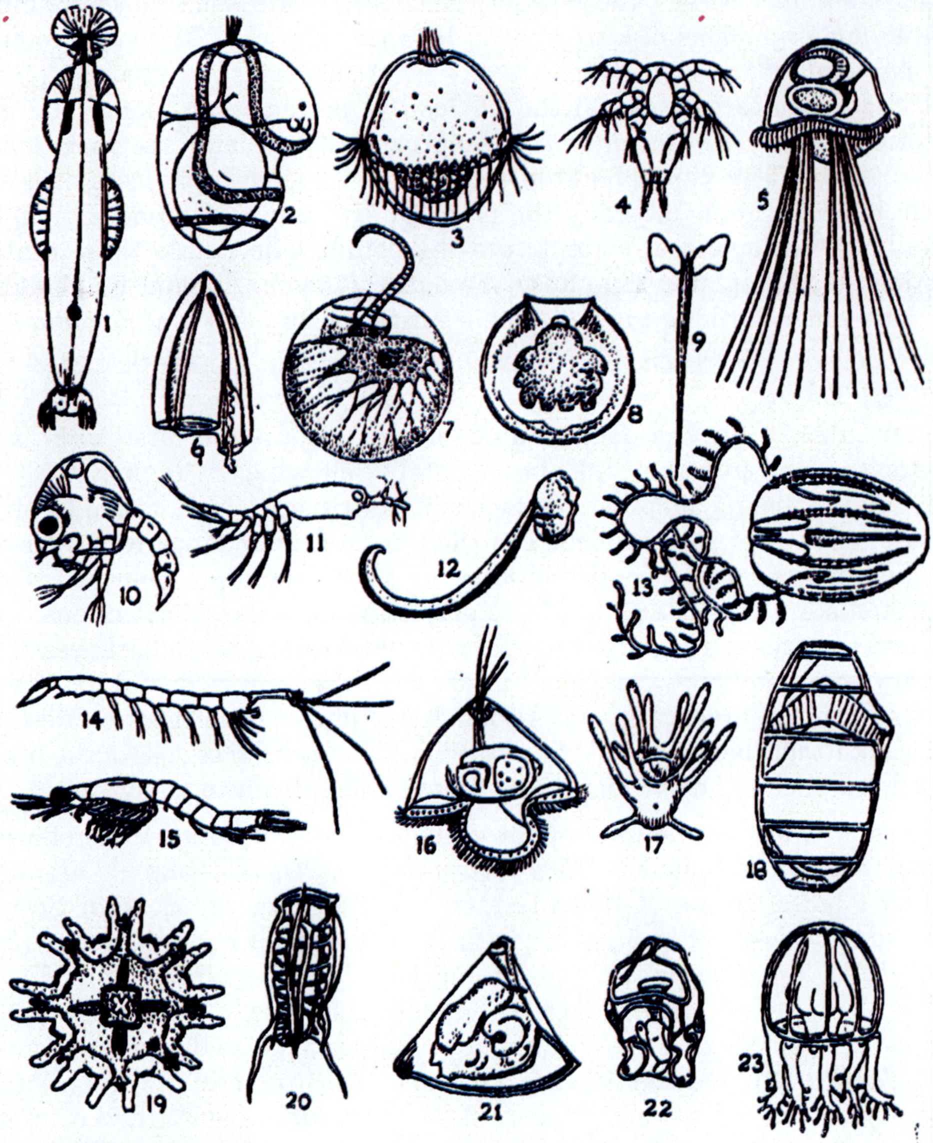

Figure 11.8. Some planktonic organisms

1. Chaetognatha, 2. tornaria larva, 3. trochophore larva, 4. nauplius larva, 5. mitraria larva, 6. siphonophore (*Diphyes*), 7. noctiluca, 8. lingula larva, 9. pteropod, 10. zoea larva 11. copepod, 12. tadpole larva, 13. ctenophore, 14. caprella, 15. mysis, 16. pilidium, 17. pluteus, 18. doliolum, 19. ephyra larva, 20. salpa, 21. cyphonautes, 22. auricularia larva, 23. cubomedusa.

Majority of the planktonic forms belong to the nannoplankton or microplankton. Special processes like spines and other structures enable many animals to maintain themselves in the environment. The horns of *Ceratium,* spines of globigerines, feathery setae of copepods, and the fleshy arms of echinoderm larvae are all examples. Different body-shapes also aid in preventing sinking. Planktonic forms assume a ribbon-like needle like or chain-like shape. Hydrozoans like *Obelia cyphonauates* larvae of the *Bryozoa* and many copepods retard their sinking by their body-shape. Additional structures enable several planktonics to save themselves from sinking. Shells when present are thin; vacuoles in the protoplasm of planktonic protozoa are useful in floatation. Droplets of oil within the protoplasm help to decrease the specific gravity (e.g. copepods and fish eggs). *Physalia* (the Portuguese-man-of war) has and air bladder (pneumatophore) which helps it to float.

Another important constituent of the pelagic realm is the nekton. Nektonic organisms are those that are capable of leading a life in the open sea by specific locomotor and swimming organs. These forms are free from the stress of wind and current. Many fishes, cephalopods, reptiles and mammals are grouped under nekton. In the oceanic realm the boundary of the deep sea may be set at 150-200 meters depth. The benthic realm lying immediately below the continental shelf is the archibenthic and that of greater depths and abyssal benthic zones. The term deep sea pertains to the latter are since those great depths possess unique features far away from the reach of sun's rays. The deep sea environment is pitch dark and this in turn affects and modifies the general nature of the environment imparting a unique identity to it. The absence of light excludes the possibility of any growth of vegetation or other photosynthetic organisms. Hence food is a major problem of the deep sea. Most of the deep sea dwellers thrive upon the bodies of surface forms that sink to the bottom while a large number of species are active predators. Some even feed on the excreta of their upper zone compatriots. Temperature being low at the abyssal zone, decomposition of the bodies falling from the above is retarded and the deep sea animals enjoy an almost fresh diet. But animals of the intermediate zones intercept the sinking food and so the deeper most inhabitants get only a small proportion of the food. The food missed by the deep sea inhabitants accumulates at the bottom constituting the food supply of deep sea sedentary forms. Echinoderms and foraminiferans are the chief dependents on such accumulated food. Predators of the deep seas bear a grotesque look with unusually large buccal cavities and a well developed dentition and possess stomachs that can be distended to any size (e.g. *Saccopharynx*). Small body size is almost an universal character of deep sea animals although gigantic forms have also been recorded.

Bioluminescence is a characteristic feature of deep sea animals. The light producing organs or photophores may be simple or complicated structures, serving specific functions especially in higher forms of animals. The light organ is situated at the tip of a movable lure in *Astronesthes* and *Melanocetus*, and the lure can be used to draw the prey towards the mouth.

The chief function of bioluminescent structure is undoubtedly to aid vision in the abyssal darkness, because in many circumstances identification of members of the same species is facilitated by the arrangement of bioluminescent organs as well as the colour specificity. Many fishes utilize this cold light to drive away enemies by a sudden flash of light.

The eyes of abyssal animals are well developed even though they live in an environment of constant darkness. In the fish *Gonostoma denudata* the eye is so large that it occupies nearly two-third of the length of the skull. But eyeless animals of the deep sea are equally common. Deep sea crustaceans have reduced eyes while cephalopods and other molluscs (e.g. *Pecten*) are eyeless. Sense organs like the antennae are well developed in deep sea animals to compensate for the absence of eyes. Extraordinarily long antennae are characteristic of these animals sometimes measuring 4-5 times the length of the body. Instances where they are 10-12 times longer the body are also reported e.g. the shrimp *Aristaeus.* Lateral line sense organs attain their zenith of development in deep sea fishes. These sense organs enable these animals to live successfully in the lightless environment in spite of the absence of organs of vision.

Abyssal animals are generally pale in colour. The average size of individuals is smaller than those of shallow water fauna, possibly because of limited food. Highly diversified fauna are present and major marine taxon is found. Benthic foraminifera followed by polychaets are numerically dominant. Isopods like *Haploniscus* are characteristic. *Neopilina*, the monoplacophoran mollusc is characteristic of the abyss. The presence of all the groups of the meiofauna (turbellarians, gastrotrichs, tradigrades, gnathostomulids, harpacticoids, kinorhynchs, nematodes and foraminifera) has been established down to depths exceeding 5000 m.

Brackish waters and Estuaries

The junctions of seas and rivers are often sites of unique ecological properties possessing a characteristic biological make-up. These sites are designated as estuaries. An estuarine condition results wherever freshwater meets the sea. The incoming freshwaters bring about an alteration of the salt content of the sea with which it is in communication. The degree to which dilution occurs is very often dependent on the size of the rivers. Estuaries typically occur at river mouths and coastal bays. Some estuarine regions possess a characteristic funnel shape and have extensive mud flats on either sides of the funnel, the mud being deposited by the tides of the sea.

A good example of the tropical 'bar built estuary' is afforded by the Cochin backwaters, because the main inlet or barmouth at Cochin supplies brackish with sea water and serves as the main mouth for other niches. Two estuarine types occur in the cochin backwaters (1) during premonsoon season and at the beginning of a monsoon, it is an estuary, almost homogeneous vertically although with varying salination (2) at the end of the monsoon and during post-monsoon, it is stratified and is continually supplied with freshwater at the surface and sea water at the bottom, so that it is a two layered estuary with vertical mixing. Organisms belonging to groups like Echinodermata, Cephalopoda, Mollusca cannot live in estuaries since they are stenohaline. Euryhaline species appear to be the most successful inhabitants of estuaries.

It must be noted that the range of salinity of water trapped in the bottom mud deposits is considerably reduced as compared to the overlying water. This water trapped in the bottom deposit is termed as interstitial water because the water is in the interstices of particles. Changes in the salinity in interstitial water takes place more slowly. This naturally allows animals living within the interstitial region to penetrate greater distances up an estuary than those subject to more rapid fluctuations of salinity as in the case of overlying water. Further the oxygen concentration is generally reduced near the bottom and the amount of dissolved oxygen depends upon temperature and salinity which varies due to the mixing of different bodies of water. These fluctuations considerably affect the colonization of estuaries by animals.

The regions of estuaries can be classified into the head, where freshwater enters the estuary, *upper*, *middle* and *lower* reaches with increasing range of salinities and the mouth with salinity nearly equal to the sea. The animals inhabiting the estuarine region are *oligohaline* (0.5 to 5 per cent), *mesohaline* (5-18 per cent) in the upper reaches, *polyhaline* (18-25 per cent) in the lower reaches, 25-30 per cent in the middle and *euhaline* (30-40 per cent) in the mouth of the estuary. Although salinity appears to be the principal factor, other conditions of the estuaries vary greatly and these changes are largely based on topography and other factors. The extent of penetration of the polychaets *Marphysa gravelyi, Diopatra variabilis, Clymene insecta, Loimia medusa, Glycera embranchiata, Onuphis eremita* into the Adayar estuary with varying salinities, is well known.

Studies on siphonophores of the estuaries in India as shown by observations in the tropical Cochin backwaters revealed their occurrence in the pre-and post monsoon period showing that salinity has a more direct role. Five species of siphonophores–*Lensia subtiloides, Lensia hotspur, Muggaea delsmani, Diphyes chamissionis* and *Eudoxides mitra* are very common in this estuary. Being a terrestrial zone between the freshwaters and seas, the estuary is an ecotone and therefore the typical estuarine forms are unique in their habits. The Hoogly-Matla estuarine system in West Bengal with an area of 3100 sq metes is the largest estuarine system of the Indian coast. A major portion of the Ganges-Brahmaputhra delta is covered by this system, wherein tidal fluctuations are considerable. The main Hoogly estuary is typically an estuary with freshwater characteristics. From the salinity point of view it falls under the mixohaline range. In any estuary, accumulation of detritus brought in by the rivers is a characteristic feature and the Hoogly estuary is no exception to this rule. Large amounts of detritus are washed ashore. There is no thermal variation; wide variations of salinities ranging from traces to 33.75 percent are noted (Gopalakrishnan, 1971). Among the planktonic forms of the estuary, several species of diatoms, *Biddulphia, Planktoniella, Hemidiscus, Chaetoceros, Cyclotella, Stephanodiscus, Triceratium*, green algae like *Pediastrum, Spirogyra, Eudorina, Tribonema, Closterium, Zygonema, Pandorina, Volvox, chlorella*, blue-green algae like *Microcystis, Oscillatoria, Anabaena, Trichodesmus*; flagellates like *Euglena, Ceratium, Peridinium, Noctiluca* are commonly observed. Among the Protozoa, *Difflugia, Arcella, Vorticella* dominate. Also common are rotifers like *Brachinus, Keratella, Asplanchna*, etc. copepods like *Diaptomus, Pseudodiaptomus, Cyclops* and *Paracalanus*, cladocerans of the genus *Bosmina, Bosminopsis, Ceriodaphnia, Monia*, besides a large number of isopods. The above mentioned forms are chiefly confined to the waters, leading a pelagic existence. Among the benthic fauna, annelids like *Tubifex*, megalopa larvae, small prawns, chironomids, gastropods, and fish larvae have been recorded.

Estuaries are excellent sites for fisheries. Many fishes migrating towards the freshwater or vice versa spend a long time in the estuaries. Endemic fishes characteristic of the estuarine environment are numerous and Gopalakrishnan (1971) has recorded from the Hoogly-Matla estuarine system the commercial fishes like *Hilsa* and other clupeids, *Harpodon*, prawns, mullets, catfishes and large numbers of palaemonid prawns. Totally about 130 species of fishes and 29-30 species of prawns are on record. Of the fishes some like *Hilsa* are typically marine in habit, but use the estuary as a nursing ground. Some like *Harpodon, Mugil, Trachyurus*, etc. migrate into the mouth of the estuary to form important fish catches, while some like palaemonid prawns are freshwater inhabitants and come to live in estuaries.

The varying salinities in different months in the Adyar estuary of the Madras coast and also at different parts at the same time present wide fluctuations in the faunal composition. The brackish water sea anemones like *Phytocoetes gangeticus, Phytocoeteopsis ramunnii, Stephensonaetis ornata, Pelocoetes exul, Boloceractis gopalai* occur in large numbers. *Acormitus flagellatus*, a jelly fish is a common inhabitant of the environment. The hermaphrodite nereid *Lycastis indica*, the tubicolous polychaet *Diopatra variabilis*, the burrowing polychaet *Marphysa gravelyi*, the oligochaete *Pontodrilus bermudensis*, the myrids *Rhopalophthalmus egregius* and *Mesopodopsis orientalis*, the crabs *Uca annulipes, U. triangularis, Neptunus pelagicus, Scylla serrata, Veruna litterata, Sesarma tetragonum*, the hermit crabs *Clibanarius olivaceus* and *C. padavenisis*, the molluscs *Ostraea arkanensis, Modiolus undulatus Meretrix casta, Cuspidaria cochinensis, Stenothyria blanfordiana, Amnicola stenothyrodies, Potamides cingulatus, Styliger gopalani, Bursatella leachii* are the most common invertebrates. Of the brackish waters of Madras fishes like *Etroplus maculatus, E. suratensis, Acentrogobius viridipunctatus, A. neilli, Panchax parvus, Aplocheilus melastigma*, etc. are very common, the last four species being perennial breeders. Besides, the mud skipper *Periophthalmus koelreuteri* is a common fish often occurring in abundance. The water snakes *Natrix piscator* and *Cerberus rhyncops* are brackish water inhabitants.

The Chilka lake on the East Coast of India in the Orissa State is the largest brackish water lake in India with an area of 906 sq.km. The channel connecting it with the sea is 35 metres long and 365 metres wide. The water in the lake is mixohaline except during the monsoon period when it becomes limnetic. Abundance of zooplankton is directly correlated with salinity, while phytoplankton is also directly correlated with salinity, while phytoplankton is also directly correlated with salinity, and inversely correlated with silica. The zooplankton and phytoplankton show abundance at different periods, the former being numerous from October-November and the latter from April-August.

The foraminiferans *Elphidium, Spirillina*, the annelids, *Nereis, Nephthys, Fabricia, Sternaspis,* few cladocerans, large numbers of ostracods, numerous copepods, a few Cumacea, some isopods like *Sphaeroma* and *Exosphaeroma*, amphipods, decapods, stomatopods, gastropods like *Thais carnifera, Cerithidia fluviatalis*, lamellibranchs like *Katelysia, Modiolus* sp. are the chief benthic representatives. An interesting isopod *Cacroleana pleioniastica* pierces and consumes the body of fishes.

Lake Chilka has contributed largely to the fishery of the East Coast of India. Waters of Chilka harbour many edible species like mullets, clupeids, perches, etc. besides prawns (*Penaeus indicus* and *Penaeus semisulcatus*), crabs (*Scylla serrata* and *Neptunus pelagicus* and oysters (*Ostrea* sp.). The primary production is maintained by the available supply of plankton in which copepods are the dominant constituents of the zooplankton and diatoms of the phytoplankton.

Mangrove Biodiversity

Coastal areas in the tropical and subtropical regions are inhabited by emergent, woody species of halophytes or land trees or bushes collectively known as mangroves which are receiving increasing attention today. They inhabit the intertidal mudflats or sand flats or coral reefs and tolerate the salinities of open sea, growing between the level of the high water spring tides and a level close to, but above mean sea level. In other words, they form essential elements of the intertidal ecosystems occurring mostly on sheltered shores penetrating into estuaries of rivers where salt water penetrates. Mangroves are easily recognized by the frequent presence of pneumatophores, proproots and viviporous seeds, the commonest species being *Avicennia* and *Rhizophora*. Mangrove forests are highly productive and are therefore of considerable economic importance, specially in timber harvesting, charcoal burning, honey, tannins, fishery resources etc. Many major productive fishing grounds are adjacent to large mangrove swamps. The organisms that support the fisheries spend part of the life cycles in these areas and are therefore natural grounds for the culture and extension of various species of marine organisms. With their increasing potential as a source for coastal aquaculture, providing immense possibilities for protein food, involving brackish water fish culture, prawn, culture, shell fish culture, and turtle farming (Fig. 11.9).

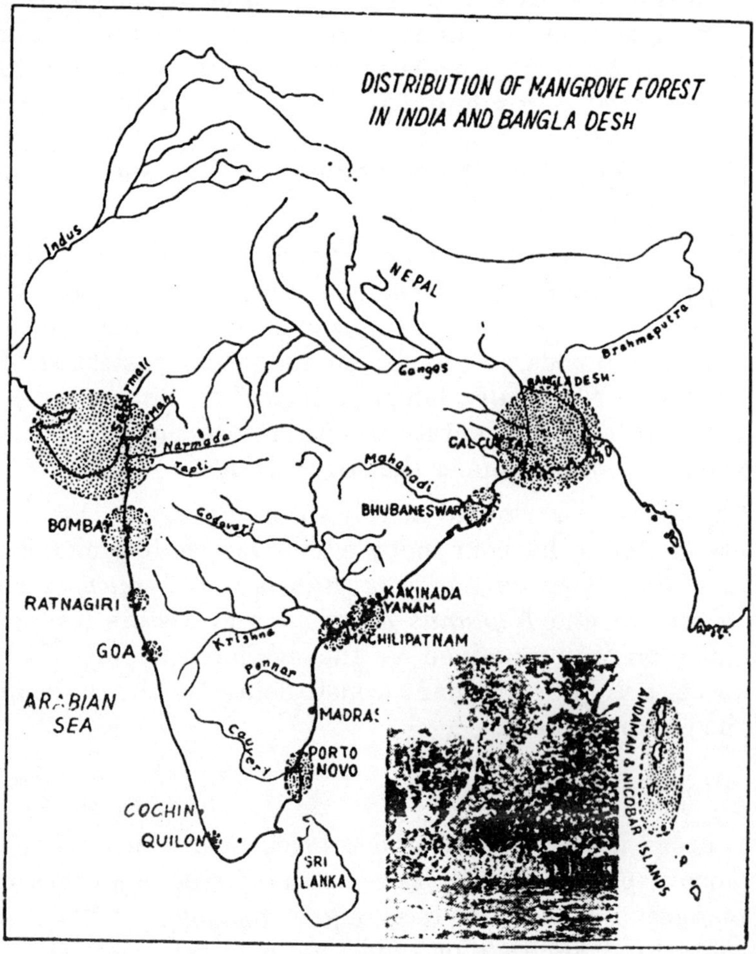

Figure 11.9. Mangroves sites in India

One of the finest examples of luxurious mangrove ecosystems relate to the Hoogly-Matla estuarine complex that achieves a good balance between physical and biotic components, resulting in a high rate of biological productivity. It consists of several basic subsystems linked together by the ebb and flow of water that is driven by the hydrological cycle (river inflow) and the tidal cycle, both of which provide 'energy subsidies' for the system as a whole. The chief subsystems are: (1) the shallow water production zones in which the rate of primary production exceeds the rate of community respirations. These include shell reefs, banks sea weed or sea-grass beds, algal mats, and salt marshes. This subsystem exports energy and nutrients to deeper water of the estuary and adjacent coastal shelf: (2) the sedimentary subsystem in the deeper channels, sounds, and lagoons; (3) the plankton and nekton, which move freely between the two fixed subsystems producing, converting and transporting nutrients and energy while responding to diurnal, tidal and seasonal periodicities; this subsystem is able to react quickly to local abundance and scarcity of available resources.

Mangroves show distinct zonations characterized by the presence of particular flora, every zone having a dominant genus and the zonation is dependent on the periodicity of inundation, salinity and soil characteristics basis into the estuarine zone with the constant influence of tides and freshwater leading to the growth of salt tolerant species. The tidal swamp forests of Sunderbans include (a) mangrove scrub involving *Ceriops roxburghiana, Avicennia alba, Aegialitis rotundifolia* along the edge of tidal water way and sheltered muddy coast, (b) mangrove forests having *Rhizophora candelaria, Kandelia rheedii, Avicennia alba, Bruguiera conjugata,* etc. at a slightly high level, (c) salt water mixed forests having *Heritiera minor, Sonnertia apetala, Acanthus ilicifolius,* etc. mostly in the largest deltas, notably the Ganges.

Mangrove environments experience (a) both minimum as well as maximum, salinity, (b) higher range of variation in silicate concentration, (c) and high range of variations of temperature. The type of mangrove system is often determined by the coarseness or fineness of substratum and though soft muddy soil is the rule in the mangrove swamps, some estuaries have also rocky shores. Since mangroves are rooted in mud there is no land substratum for sublittoral fringe organisms and so under the shade of mangroves occur a community of organisms including alphaeids and polychaets, and on the mud surface are found *Clibanarius, Pseudopagurus,* hermit crabs, mud skippers like *Periophthalmus*, etc.

In the mangroves in southern India are found typical littorinid molluscs, with *Littorina seaba, L. intermedia* and *L. undulata* on leaves, flowers and branches and *L. melansotoma* on the leaves of mangroves. Occurring abundantly on the leaves, trunks and branches are *Assiminea brevicula* and *Neritina depneura*. Wood-borers inhabiting roots, trunks and submerged branches are *Martesia striata, Bankia edmondsomi, Bankia campanulata* and *Teredo juttingae*. Bruguiera forests below the landwood fringe, with firm soil littered with twigs and branches, harbour a large variety of animals like the crabs *Sarmatium*, *Illyograpsus*, *Sesarma*, *Metopograpsus*, *Cleistoroma*, *Tylodiplax*, *Macrophthalmus*, etc. In rhizophora forests where also the soil is firm and compact, are alphaeid shrimps, the blue crab *Metopograpsus latifoons* and *Telescopium telescopium* among the larger snails. In *Avicennia* and sometimes *Sonnertia* shores, with soft mud, are found burrowing forms and the mudskippers *Boleophthalmus*, *Periophthalmus* and *Periophthalmodon*. A good representative of mangroves is *Scylla sonata* the portunid crab and the burrowing anomurans *Upogebia* and *Thalassina*.

Table 1. Occurrence and use of economically important prawns of Sundarban Bay Islands mangroves (After Das and Nandi, 1999)

Species	Sundarbans	Occurrence	Utilisation Bay Islands
Palaemon (Exopalaemon) styliferus	Abundant	—	Edible, Fishmeal (Juveniles)
Macrobrachium lamarrei	Common	—	Edible, Fishmeal
Macrobrachium rosenbergii	Occasional	—	Edible
Metapenaeus affinis	Abundant	Occasional	Fishmeal
Metapenaeus brevicornis	Abundant	—	Edible
Metapenaeus dobsoni	Rare	Abundant	Edible, Fishmeal
Metapenaeus lysianasa	Abundant	—	Fishmeal
Metapenaeus ensis	—	Common	Edible, Fishmeal
Metapenaeus monoceros	Abundant	—	Fishmeal
Parapenaeopsis sculptilis	Abundant	Abundant	Edible
Penaeus monodon	Abundant	Occasional	Edible
Penaeus japonicus	Occasional	—	Edible, Fishmeal
Penaeus merguiensis	Common	Abundant	Edible, Fishmeal
Penaeus semisulcatus	Common	Occasional	Edible
Acetes species	Common	Common	Edible, Fishmeal

Table 2. Occurrence and use of commercially important crabs of Sundarban and Bay Island (After Das and Nandi, 1999).

Species	Occurrence		Utilization
	Sundarbans	Bay Islands	
Scylla serrata	Abundant	Abundant	Edible
Portunus pelagicus	Occasional	Common	Edible
Portunus sanguinolentus	Occasional	Common	Edible
Charybdis (Charybdis) feriata	Occasional	Occasional	Edible, Fishmeal
Charybdis (Charybdis) rostrata	Abundant	Rare	Fishmeal
Charybdis (Charybdis) helleri	Rare	Common	Edible, Fishmeal
Charybdis (Charybdis) orientalis	Rare	Common	Edible
Thalamita crenata	—	Common	Edible
Thalamita prymna	—	Common	Edible
Matuta planipes	Abundant	Common	Edible, Fishmeal
Matuta lunaris	Common	Common	Edible, Fishmeal
Matuta banksii	—	Common	Edible
Calappa lophos	Occasional	Occasional	Fishmeal
Calappa pustulosa	Occasional	Occasional	Fishmeal
Varuna litterata	Abundant	—	Edible

12,000 ha of mangroves were cleared in Little Andaman mainly for fuelwood and this had detrimental effects on offshore fisheries.

Major threat to crustacean diversity is due to over-exploitation and collection of undersized specimens as well as large scale exploitation of prawn seeds. Over-exploitation is also likely to have an adverse effect on the population of commercially important species. According to a report, to catch 8,106 tiger prawn seeds in Sundarbans, collectors destroyed juveniles of 1,304,937 other prawns, 60581 fishes, 246097 crabs, 8391 molluscs and 66792 unidentified meroplanktons (Das and Nandi, 1999). Often, many species are harvested indiscriminately without knowing the effects of over-exploitation on the species and the ecosystem.

Ten species of crabs in Sundarban mangroves, such as *Rhynchoplax woodmasoni, Rhynchoplax inacechoides, Hymenicoides carteri, Uca lactea, Pachyagrapsus porpinquus, Ptychognathus onyx, Pyxidognathus deianira, Sesamoides longipes, Neosarmatium smith* and *Clistocoeloma merguiense* and 3 species in mangroves of Bay Island, namely, *Nanosesarma andersoni, Metaesarma rousseauxii* and *Leipocten sordidulum* are vulnerable/rare species, besides 5 species of xanthoid crabs, 2 pilumnids (*Pilumnopeus indica* and *Heteropilumnus ciliatus*) and one menipid crab (*Myomenippe hardwicki*) are recorded as rare in Hoogly-Matla estuary of deltatic Sudundarban. However, during 1980's *Myomenippe hardwicki* was found to be common in the reclaimed and forested parts of Sundarban (at Canning and Saimari areas).

CHAPTER - 12

FRESH WATER BIODIVERSITY

Introduction

The inland freshwaters encompass a diverse array of ecosystems as varied as lakes and rivers, ponds and streams, temporary puddles, thermal springs and even pools of water that collect in the leaf axils of certain plants. This is a small fraction of world's water resource. Despite this, inland aquatic habitats show far more variety in their physical and chemical characteristics than marine habitats and contain a disproportionately high fraction of the world's biodiversity. Fresh water organisms originated from marine ancestors. Insects, hydracharina and pulmonate snails for instance owe their freshwater habits to the terrestrial environment.

Inland water habitats can be classified into stagnant (*lentic*) and flowing (*lotic*). They may also be classified into perennial or transient. Each of these has its own set of distinctive ecology and biological community. Lentic systems comprise lakes and ponds. Majority of large lakes are formed either by glacial or tectonic activity. Most of the glacial lakes are geologically young and were formed during Holocene, 11,500 years ago. Only about 10 existing lakes are known to be present much before Holocene. These Old lakes such as lake Baikal and Tanganyika are biologically very interesting and tend to be species rich. Man-made lentic habitats such as irrigation tanks, ponds and reservoirs are predominant landscape features in some regions such as Deccan in peninsular India.

Lotic systems encompass rivers and streams. A river system is essentially a linear body of water draining under the influence of gravity. Most of the river systems discharge into the sea; some into lakes, such as the river Vaigai in the Western Ghats, which emplies itself into the "Ramanad big tank" in the Ramanad district of Tamil Nadu. A few watercourses in arid regions enter inland basins where no permanent lakes exist and disappear into the dry plains.

Large rivers such as Ganges and Brahmaputra cross over many degrees of latitude and traverse a wide range of climatic conditions. Variations in water flow and underlying geology also create a wide range of habitats, often within a short distance. Because of this change in habitats, different organisms are typically present in different parts of any given river system. Even though rivers are physically very dynamic, large rivers rarely disappear, and there are indications that some of the large rivers are in existence for tens of millions of years. This is reflected in the fact that, all the taxonomic groups are found in running waters, and some invertebrate taxa are exclusive or attain greatest diversity there.

Biodiversity at higher taxonomic levels in freshwater systems are much narrower than those in the terrestrial or marine systems. Only one extant eukaryote phylum (Gamophyta-green conjugating algae) is confined to freshwater habitats. The overall number of species (species richness) is also low compared to marine and terrestrial groups. However, species richness in relation to habitat extent may be very high. For example, about 10,000 (40%) of the 25,000 known fish species are freshwater forms. This high diversity of freshwater systems fishes relative to habitat extent is promoted by isolation between freshwater systems. The species richness in the freshwater systems increases towards the equator as is the case with terrestrial habitats. There are many more species in the tropical freshwater systems than in temperate regions, but in some specific groups such as freshwater crayfishes this trend appears to be reserved.

Fresh Water Plant Diversity

About 5,000 species of large algae from the groups the green, brown and red algae are known. A major proportion of these are marine or brackish. The green algae Chlorophyta include order Ulotrichales have about 80 freshwater species. However, the 440 species of stoneworts are almost entirely found in freshwater.

Freshwater fungi are poorly documented when compared to other taxa, especially from the tropics. More than 600 species are currently recorded, more from temperate regions than from the tropics. Most of these fungi are Ascomycetes with few Basidiomycetes and Zygomycetes. They use vascular plants as their substrate. Freshwater fungi are important as parasites endophytes and saprophytes of aquatic macrophytes; as decomposers of woody debris and as food source for invertebrates.

Liverworts and mosses represent non-vascular plants of freshwater system. Many of the hygrophilous terrestrial mosses (e.g. *Thamnium*, *Bryum*,

Mnium) have aquatic forms. The family Ricciaceae commonly known as liverworts have a number of species growing submerged in freshwater bodies. The ferns (Filicinophyta) representing lower vascular plants are very diverse in freshwater systems. Certain aquatic pteridophytes such as *Salvinia* sp. (*Salviniaceae*) are noxious weeds; others like the members of Azollaceae support symbiotic nitrogen-fixing bacteria *Anabaena azolla* (*Cyanobacteria*).

Most of the aquatic plants species are relatively widespread and many are cosmopolitan. It has been estimated that about 1 per cent of angiosperms i.e., about 2,500 species are aquatic. The family Podostemaceae is of particular interest for its many monotypic genera and narrow endemics. Most of these endemics are restricted to a single river.

Freshwater Animal Diversity

Animal species are far more diverse and numerous in inland waters than plants. Apart from fishes, invertebrates from an important group. The important groups include sponges, flatworms, molluscs, polychaete worms, oligochaete worms, crustaceans, insects and numerous parasite species in various groups. As on land, insects are the most diverse group of organisms in inland water. Unlike terrestrial faunas, where beetles are more abundant, the (Diptera) appear to be by far, the most abundant group in inland waters.

Vertebrates of freshwater systems are very diverse. Of all freshwater vertebrates, fishes are the most well known and valued. About 10,000 species of fishes are confined to freshwater. This is about 45% of known fish species. Another 1100 or so species occur in freshwaters but they are not confined to it. A significant proportion of freshwater fishes are from just four orders; the carps and their relatives (Cypriniformes); the characins (Characiniformes); the catfishes (Siluriformes); and the perches and their relatives (Perciformes). The first three of these are entirely freshwater forms, while the last has a significant number of species both in marine and freshwater systems.

Amphibians, comprising about 5,000 extant species have aquatic larval stages and wholly depend on freshwater for survival. Since 1990, there is worldwide decline in amphibian populations and species disappearances. Various factors have been attributed for this alarming trend, but the single factor that has been of worldwide importance is habitat loss. Crocodiles, water snakes, turtles and terrapins represent aquatic reptiles. About 228 species are found in various freshwater ecosystems. Most of the reptiles are semi aquatic and they come to land to reproduce.

Birds are the most spectacular of all freshwater organisms, though none of them are wholly aquatic. Like reptiles, all the water bird species lay eggs on land. They are equally abundant and diverse as freshwater fishes. Nearly 250 species are exclusive to freshwater habitats. A large proportion (60%) of this is form a single family, Anatidae comprising ducks, geese and swans. It is interesting to note that, the Passeriformes (perching birds) that dominate terrestrial systems in terms of species richness is poorly represented in freshwater systems. Only one passerine family Cinclidae (dippers) can be considered truly freshwater in its habits. A large number of wading birds such as sandpipers (Scolopacidae), plovers (Charadridae) herons and bitterns (Aedeidae) and other diverse assemblage of families are associated with inland freshwater habitats. They are facultative to freshwater habitats, as they also use coastal wetlands. They form the bulk of the water fowl. A small number of birds such as kingfishers (Alcedinidae, Dacelonidae and Cerylidae), fish-owls (Strigidae) and fish-eagles (Accipitridae) have evolved to dive from the air to the water surface and feed on fishes and other organisms.

The Cetacea is represented in freshwater by five species of river dolphins (Platanistidae) and the Sirenia by three species of manatees (Trichechidae). About another 65 species of mammals are also found in inland waters and they lead more or less an amphibious life. Nearly 50% (32 species) of this form the aquatic mice (Muridae). Other well known inland aquatic mammals include the platypus (Ornithorhynchidae), Hippopotamus (Hippopotamidae), beavers (Castoridae) and otter (Mustelidae).

Freshwater Biodiversity Hotspots

Spatial extent of the distribution of freshwater organisms is limited by the availability of suitable habitats. The distribution of most of the freshwater species especially fishes, molluscs and crustaceans is limited to present or formerly continuous river basins or lakes. The principal barrier for the distribution of species is the watershed between the systems. Based on the available information, a recent analysis recognized 24 freshwater biodiversity hotspots. This analysis was based on the diversity of freshwater molluscs, crabs, fairy, shrimps, crayfishes and fishes. It is interesting to note that, freshwater biodiversity hotspots more or less match with the biodiversity hotspots prioritized based on terrestrial organisms.

Human Significance of Freshwater Biodiversity

Freshwater is essential for the survival of mankind, both for drinking and for the food production. Other uses include transport, industrial production, cleaning, waste disposal, power generation and recreation. In populous developing countries such as India, the limited availability, uneven distribution and growing demand makes freshwater a precious resource. About 70% of freshwater used are conflicting and create tremendous pressure on freshwater resources especially on freshwater biodiversity.

Primary use of freshwater species is as food. Wildly caught fish from freshwater habitats is the important source of protein, especially for the communities of developing countries. In addition to food, a substantial number of species are used as medicine, fiber and fertilizer. Many more species, particularly fishes are semi-domesticated and widely used for ornamental purpose. Inland capture fisheries form a staple part of diet particularly in land locked countries and nearly 75% of protein requirement. Freshwater molluscs, crayfishes and shrimps are also exploited for food. Other groups of animals used in minor quantities for food include frogs, turtles, waterfowls, otters and beavers.

Rice, the most important food crop of the world is a produce of freshwater ecosystem. Still a large number of wild relatives of rice species are found in freshwater habitats. These wild relatives of rice species are a source of genetic Conservation of freshwater habitats is essential for the survival of these wild relatives. In addition to rice, members of the edible aroid (Araceae) such as *Colocasia* and giant swarp taro (*Crotosperma chamissionis*) form important food crops in Africa, Asia and Caribbean Islands. A large number of aquatic plants are important in local health traditions for example *Nelumbo nucifera* and *Rotula aquatica* are commonly used in Ayurvedic health system in India.

Threats and Current Status of Freshwater Biodiversity

Physical alteration and destruction of the habitat is the single major cause for freshwater biodiversity decline. Diversion and containment of rivers is known in tropical Asia for centuries. For examples, Kaveri Delta canals in India were constructed during second century A.D. Explosive population growth in tropical Asia has caused significant environmental problems in drainage basins. Deforestation leads to an increase in sediment-rich surface runoff and stream flood flow. This adversely affects freshwater organisms, especially fishes, as for example, fish decline in sediment load and decreased food availability due to deforestation. Destruction, habitat alteration and creation of new habitats by dams and reservoirs are the major threats to freshwater biodiversity. Dams, especially the large ones drastically affect the flow regimes. This can catastrophically disrupt the life cycles of species that migrate up and down the rivers.

It is estimated that at least 20% of the world's freshwater fish fauna is already extinct or on the verge of extinction. A recent study on the freshwater fishes of India shows that 59 species were seriously threatened as result of human activities. Alteration in natural flow regimes are damaging to fish stocks. For example, flood control embankment in the Jamuna River in Bangladesh have reduced the fish stocks by 50% since 1983. After the construction of Farakka Barrage across the Ganges in 1975, the *Hilsa ilisha* population dwindled to virtually nothing. A wide range of non-native species have negative impact on freshwater biodiversity. These include plants (*Ipomoea fistulosa*), snail (*Ampullaria gigas*) shrimp (*Macrobrachium nipponense*), Amercian bullfrog (*Rana catesbiana*) and fishes (*Cyprinus carpio, Pseudorasbora parva, Gambusia holbrooki, Peocillia reticulata* and *Oreochromis mosambicus*). The artificial stocking of grass carp (*Ctenopharyngodon idellus*) caused the virtual disappearance of submerged macrophytes and caused algal blooms.

Apart from invertebrates and fishes, other vertebrates such as amphibians, reptiles, birds and mammals are also affected by deterioration of freshwater ecosystems. A recently concluded study by the Bird Life International (2001) estimated that 20% threatened birds of Asia depends on wetland habitats. These include highly threatened birds such as white-winged duck (*Cairina personata*). Of the other threatened vertebrates excluding birds in Asia, include: the crocodiles, river dolphins, otters, the swamp deer, the Indian rhinoceros and the brown antlered deer.

This ecological vulnerability of peninsular Indian rivers is alarming, as a large number of people depend upon these rivers for their entire freshwater requirements. There is an urgent need to intervene at this juncture and design mitigation measures to conserve the riverine ecosystems of the region. One of the first and significant step towards this goal is the systematic scientific documentation of biodiversity of these riverine ecosystems. Such documentation, along with a record of its ongoing changes is essential to develop a conservation and monitoring programme for the riverine ecosystems.

Classification of Freshwater Habitats

Both lentic and lotic environments are in a state of gradual evolution since a lake can become a pond and finally a dry land depending on the non-availability of water or a tiny rivulet may gradually widen and deepen into a river. The fauna of these environments have to withstand this evolutionary modification of the habitat by either evolving along with the environment or adapting themselves to the stress of alteration. If not, they are eliminated from their environments. Adaptations of evolutionary significance being conspicuous in lotic waters, a consideration of the characteristics of running waters and their faunal composition may be in order.

Running waters: Biotic Environment

Biologically the water current of the lotic water are of considerable significance for the fauna. The fast moving stream has two distinct, yet interrelated habitats influencing each other–the riffle region and the pool region. Although the two regions can be distinguished by many conspicuous features, they are unique in having a uniform faunal composition. There is a mutual exchange of fauna between the two habitats. Riffles seem to harbour a richer fauna especially on the bottom of stones and pebbles. Mayfly and caddisfly nymphs and stone flies live here. Damsel flies and dragon flies along with mayfly nymphs take to life in pools. The current of water carrying abundant food materials is the most important factor contributing to the richness of fauna in the riffles. The rich oxygenation facilitated by the water movement acid in respiration. The animals living in fast moving torrential streams may be grouped as follows (Fig. 12.1).

Animals of Pools and Watersides

This region is characterized by a gentle flow of water. The dead organisms brought by the flowing water get isolated by the gentle current at the sides enabling them to settle down and form the food for the pool dwellers.

Burrowing Animals

The burrowing animals are blessed in many ways. First they live in surroundings rich in oxygen and food. Second through burrowing they are saved from the attack of predators and enemies. The chief inhabitants of the region are the dragonfly larvae (*Gomphinae*) and mayfly larvae (*Palingenia and Polymitarcys*).

Swimming Animals

Majority of animals swim with the tide and very few swim against the current. These animals swim for a long time and settle into the deep pools becoming inhabitants of the pool region at the sides of the streams.

Swift Current Inhabitants

These are the permanent dwellers of torrential streams and are perfectly adapted for a life in such a habitat. They can withstand the abrasive action of the current and are capable of streaming the current. Among the principal subdivisions of the category are:

(1) *Plant inhabiting forms*: These animals live among the mosses and flowering plants like *Eriocaulon miserum, Hydolyum lichenoides* and *Duroea wallichi*. Animals living here have a torpedo like body. This type of body enables them to offer the minimum resistance to the current. Some live entangled in the roots (funnel-mouthed tadpoles of *Megalophrys* and tipulid larvae). They possess devices for fixation and have hydrostatic organs.

Some others live among leaves and stems (*Cephalopteryx, Herlodes, Phalacrocera, Gammarus* etc.). These animals possess hook-like structures on their body helping them in anchoring to leaves and stems. But the larvae of simulids and chironomids live on the exposed surface of plants.

(2) *The Rock Inhabitants*: Surface of rocks exposed above the level of water support a number of animals. Among these, some like *baetis* larvae cling by means of claws, while others like the larvae of *Simulium* and chironomids achieves this by grappling hooks at the posterior end of the body. There are a few (caddisfly worms) which posses both these structures. They possess a torpedo-like body. The mayfly nymphs *Iron* and *Psephenus* attach themselves to the rocks by means of functional pads. The undersides of rocks similarly harbour a variety of animals and like the larvae of stoneflies. They are all uniformly flattened and adhere by applying their ventral surface to the rocks (Fig. 12.2).

The *torrential stream inhabitants* in general are characterized by a streamlined body. This helps in reducing the resistance to the medium as in larvae of mayflies like *Baetis*. The chief inhabitants of torrential streams are those which are exclusively adapted for that mode of life and the complete absence of plankton is a characteristic feature of the environment. A few ciliates have been recorded although they show no special adaptations for this mode life. They live at the bottom of the stream. The velocity of the current becomes gradually reduced as depth increases and after a certain depth the velocity is almost zero. The ciliates are found living here and the recorded species are similar to those of still waters. Plecopteran nymphs (*Perla*) have elongated and flattened bodies and possess movably articulated paired claws. The nymphs also possess a series of backwardly directed spines in the various regions of the body and these enable them to maintain their position in narrow crevices among stones. Larvae of the odonate families Megapodagrioninae, Eupallaginae and Protoneurinae apart from Gomphinae and Libellulinae are waterfall dwellers of India. Of these, those that live on the substratum are flattened dorsoventrally, while others that grapple on the rocks by means of their claws and swimming with the current have a cylindrical form. The principal ephemeropteran families are Heptagenidae, Beatidae and Prosopistomidae. The eggs of these groups have devices like the micropylar structures or threadlike chorionic extensions helping them to anchor to the substratum. Nymphs of Heptageniidae are flattened dorsoventrally with a broad and ellipsoid head. They have robust legs which are flattened, and their terminal claws are powerful. Certain forms have their gill lamellae modified for attachment. They possess well developed anal cerci. The commonest genera of Indian waters are *Ecdyonurus, Epeorus and Iron.* Baetidae comprise active forms and are adapted to maintain themselves at the tips of the falls. *Prosopistomida* represented by *Prospistoma* have the head and body broadly oval and great depressed. They have a convex dorsal surface and a concavae ventral surface. The body is margined by a strong and narrow chitinous band which is closely applied to the substratum. The surface of the band is covered with small papillae on the ventral side which help the animal in adhesion. Among the Heteroptera, only two types of naucorids are reported to live in torrential streams. They attach themselves by pads on the tibia or by claws (Fig. 12.3).

Figure 12.1. Stream Profile (Courtesy, Krishnan and Sivaramakrishnan)

Figure 12.2. Rocky substratum inhabitants (Courtesy, Krishnan and Sivaramakrishnan)

From among the Trichoptera two types may be recognized, free living and shelter building. In these types the abdominal hooks and claws are the chief organs of fixation. Shelter forms may build shelters which are permanent or which are portable. The fixed shelters have been found effective in the environment. Certain forms live under rocks by building a barricade of pebbles, the latter being held together by a silken secretion. Some shelters are of thin spun silk covered with mud or slit. A typical example is the genus *Philopotamus*. Portable cases are dorsoventrally flattened, the upper surface is slightly arched and the under surface flattened. The cases are formed of small bits of stones and are therefore heavy. Some tubes are elongate and rectangular. Lepidopteran larvae of the genera *Aulacodes, Elphila* etc. live in rocks in swift currents under thin films of silk. The animals stem the flow of current by means of crochets on the forelegs and terminal claws on the thoracic legs.

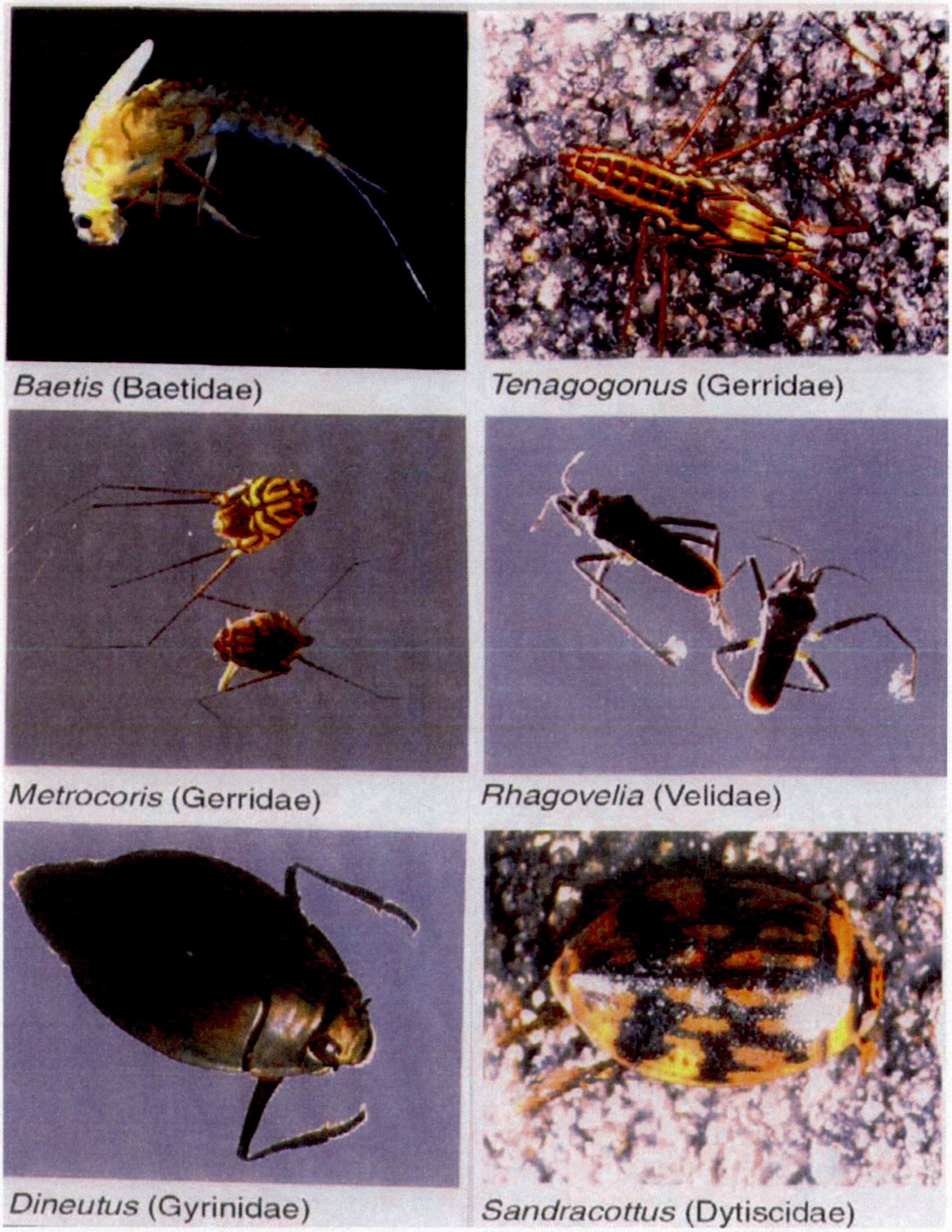

Figure 12.3. Running stream fauna (Courtesy, Krishnan and Sivaramakrishnan)

Figure 12.4. Some niche-specific freshwater insects
(Courtesy, Krishnan and Sivaramakrishnan)

The coleopteran genera *Psephenus* and *Psephenoides* abundantly represented in India, do not possess any adhesive devices, but have slender legs armed with claws. Among the Diptera the psychodid larvae of *Periocoma and Ulomiyia* live within mosses. The backwardly directed processes of their body help in entangling them to the plant and prevent them from slipping. The larvae of *Maruina* are dorsoventrally flattened and are provided with a row of light suckers on the ventral flattened surface. The pupae are cemented to the rock. The tipulid larvae (*Triogma* and *Phalacrocera*) anchor themselves by their leaf-like long processes which are denticulate and serrate in various ways. The larvae of *Tipula* live among the roots of plants where they are not affected by currents while those of *Diceranota* and *Pedicia* are provided with pseudopods which end in rows of chitinous hooks. The larvae of *Antoclia* live under the shelter of thin spun silk and the pupa fixes itself to the case with two powerful hooks at the caudal end and lies in the current with its head pointing downstream. The chief means of attachment in the chironomids are the hooks on the posterior pair of pseudopods. The tubes characteristic of the standing water forms are dispensed with and certain chironomids like the larvae of *Cardiocladus* and *Charadromyia* live under sheets of thin spun silk. The larvae of *Simulium* attach themselves by means of salivary secretion and hooks on the posterior appendages. In Blepharoceridae suckers are the chief organs of attachment. The larvae of *Deuterophlebia* possess seven pairs of grappling organs believed to act as suckers. They hold on to the substratum by means of claws (Fig. 12.4).

Torrential fishes are successfully adapted for life in the extreme environment. Many of them are capable of moving against the currents or waterfalls. Reduction of resistance is brought about by a fall in size and the assumption of a streamlined body. Some of them like the trout and mahseer achieve this movement especially in rapid and slow falls by muscular effort. But when a fall is more than 20 to 30 feet high it becomes a barrier for the migrating fish. Behind such high falls, there are wet rocks constantly moistened by the spray of water or water trickling from the rocks above. Certain fishes provided with special organs of attachment crawl up such wet surfaces. But the majority of torrential fishes live on or among rooks, and boulders, over which water flows very swiftly. The fishes of these habitats have to put up with much less force of rushing water unlike those that live on the rocks which have to bear the brunt of the current. Accordingly small loaches (*Noemacheilus*) or loach-like fishes (*Ambyceps olyra*) are met with at the bottom and limpet-like fishes (e.g. *Glyptosternum*, *Balitora*) or fishes with special modifications (*Garra, Glyptothorax, Pseudecheneis*) are seen on the rocks. The respiratory organs of torrential fishes have also undergone structural modifications to suit the high oxygenation of the surrounding water. The gill openings become restricted and the gills themselves are reduced so that such fishes cannot live long in sluggish waters poor in oxygen. The loaches and the carps adhere to rocks like the limpets, with the ventral surface, inclusive of paired fins forming on large sucker, e.g. *Gastromyzon* of Borneo, the Indian loach *Balitora* of the Eastern Himalayas and the Assam hills (Hora, 1947; Hora and Law, 1942) *Bhavania* and *Travancoria* are also typical torrential stream inhabitants. Besides, a number of amphibian tadpoles live in the environment. In Batrachia, adhesion is effected by hypertrophy of the lips which on the ventral aspect are covered with teeth or papillae. Haunting the most turbulent of small hill streams of Himalayas are a number of birds like *Chaimarrornis leucophala* or the white capped red start, *Rhyacornis fuliginosa, Myophonus coerules*, the Himalayan whistling thrush and *Miclocichla scouleri*, the little fork-tail.

Large amounts of detritus present often attracts a large number of detritus feeders and ponds where tube inhabiting annelids are especially abundant. They secure their food through their attachment anterior ends. Many burrowing may be superficial, resembling grazing as in *Sialis* (alder-fly) or deeper, extending to great depths as in midge *Chironomus* and branched burrows in mayflies. Among the molluscs are *Sphaenius, Psidium* and *Anodonta dominata*. A number of bryozoans are also present especially on the shells of snails. Large numbers of fishes thrive in slow streams due to the sumptuous food supply. Chief among these fishes are carps, cat fishes, suckers, stingers and spoonbills. The richness of nourishment afforded by slow streams attracts a large number of vertebrates like crocodiles and river turtles, apart from amphibians.

The Lentic Environment

Lentic environments have been extensively investigated mostly due to the stability of their flora and fauna facilitating long time observations. Lakes and ponds besides swamps and impoundments are categorized under lentic environments. Any standing body of the water can be classified into a trophogenic zone, and a *tropholytic zone* on the basis of the depth of light penetration enabling photosynthesis. The former is often distinguished by the abundant plant growth and dependent fauna, while the later denotes a general absence of vegetation and habours mostly saprobes. In between the two zones is the compensation depth which is also recognized as the boundary between the two zones. Marking out the compensation depth is the perfect equilibrium exhibited in the process of respiration and photosynthesis. In other words, energy input and output balance each other. The trophogenic zone is further classified into the littoral zone or the *pedonic zone* where water is shallow and light penetrates up to the bottom, and the *limnetic zone* which is the vast expense of water that stretches up to the zone of effective light penetration. The term *benthic zone* is applied to the bottom of a lake or a pond and the organisms living in the benthic zone constitute the *benthos*.

Lakes and Ponds

Environmentally, lakes are totally different from the running waters. Apart from the absence of currents, a distinguishing feature of flowing waters, lakes show conspicuous variations of various physico-chemical factors with the changing day, year or season. The thermal and oxygen stratifications resulting in a themocline and oxycline are especially important. But lakes of high altitudes rarely present conspicuous temperature stratifications (Mani, 1968c). This phenomenon can be distinguished by the presence of a very high oxycline. Various classifications of lakes on the basis of physical factors, productivity, etc. exist. Based on temperature Hutchinson (1957) classifies lakes into *dimictic, monomictic* and *polymictic* lakes. The dimictic lakes exhibit two overturns every year, while monomictic lakes are further classified into *cold monomictic* and *warm monomictic*, the former being characterized by a circulation only during the summer, while the latter has a complete circulation in winter as well. Polymictic lakes present circulations throughout the year. Based on productivity, lakes may be *oligotrophic* when productivity is low, or *eutrophic* when primary production is very high. Oligotrophic lakes present scanty to dismal faunal composition and yet are more stable in nature. Eutrophic lakes are characterized by a rich fauna, and a good amount of variation is their physico-geographical features mark them out from the oligotrophic lakes.

Impoundment is an artificial man-made lake. It may be a close type or an open type depending on the discharge of water into it. Sandynulla is a closed impoundment, while pykara is an open impoundment. Impoundments differ from natural lakes in thermal and oxygen variations and are often characterized by a low percentage of benthos.

Ponds are lentic habitats with a relatively small area but with greater depths than lakes. Ponds are typical examples of ecosystems with all the characteristic trophic levels and energy transfers discussed earlier. The pond ecosystem is illustrated in Fig. Temporary ponds, which are very often shallow and quick drying, present an extreme type of lentic environment. Various organisms of the lentic environment can be ecologically classified based on whether they are dependent on the substratum or free from it. Organisms depending on the substratum are called *pedonic forms* and those that are free from it are the *limnetic forms*.

The greatest diversity and abundance of lake fauna is encountered in the littoral region or pedonic region especially up to the depth to which effective light penetration is possible facilitating the growth of rooted vegetation. The plants as well as the lake bottom in this region harbour a rich fauna. The oxygen produced by the green plants as well as shelter and protection they afford help them to thrive well. With an abundance of emergent vegetation having an admixture of swamp plants the littoral zone as it gradually slopes down has a large number of plants with floating leaves. Beyond the zone of floating vegetation is the region occupied chiefly by submergent vegetation. In lakes, where wind stirred waters constantly strike against the shore, vegetation naturally appears scanty. But ponds usually have a rich flora, due to the greater depths and lack of disturbance.

Emergent plants of the shoreline possess roots, but their tops with thin chlorophyll bearing regions are exposed. Interspersed with these forms are the cat tails, wild rice, arrow heads, etc. Floating plants remain at the water surface, but rooted to the bottom. Characteristic plants of this region are *Nymphaea, Potamageton*, etc. The zone of submerged plants like *Elodea, Vallisneria, Myriophyllum*, etc. are rooted at the bottom but remain beneath the water surface.

Limnetic zone is the region of rapid variation, with the water level, temperature and oxygen content varying from time to time. In mountain lakes which are glacial in origin, as in the Himalayas, change in water level is a conspicuous phenomenon, the onset of summer bringing forth a rapid and sudden rise in water level. Various protozoans which are capable of encystment, tardigrades like *Macrobiotus*, rotifers like *Rotaria, Philodina,* copepods, besides snails and frogs occupy this region. All the above groups can successfully withstand desiccation when the shores dry up in summer due to the fall in water level but become active with the onset of the rainy season. Sedentary and slow moving forms are excluded from this environment due to the constant threat posed by predators and the absence of a permanent substratum for attachment. The emergent vegetation has a rich and diverse fauna existing on, above, and near it. Protozoans, annelids and rotifers are numerous; besides the vast number of crustaceans. Molluscs especially snails and mussels are abundant. Insects and insect larvae (damsel fly and dragon fly) thrive well here in the conductive habitat afforded by the luxuriant plant growth. The submerged vegetation has the freshwater bryozoans and other sessile forms.

With the exception of a few rotifers like *Keratella, Brachionus, Diurella,* and *Trichocerca* and crustaceans like *Cyclops, Daphnia,* and *Moina*, the majority of the freshwater animals of standing waters depend in one way or another on the vegetation. Although free swimming forms like *Ranatra, Corixa, Dytiscus,* larvae of *Culex* and of *Chaoborus* are free from any attachment, some insects like *Gerris* and *Gyrinus* are found swimming on the surface of waters. Many insects and molluscs are found on the vegetation. Larvae of *Dytiscus, Laccotrephes*, dragon fly nymphs, caddisfly larvae and leeches like *Glossophonia* and molluscs like *Lymnaea* are found only on vegetation. Larvae of *Chironomus* are found to inhibit the interior of rotting vegetation. But planarians are found underneath the leaves of floating plants. The bottom of the lentic environment has an abundant fauna, chiefly constituted by detritus feeders. Besides, there are many forms that live feeding on the stems and leaves of submerged plants. Insects, mites, amphipods, and pulmonate snails depend on these plants. Protozoans like *Vorticella, Stentor,* coelenterates like *Hydra* and rotifers like *Brachionus* are also attached to the undersurface of leaves (Fig. 12.5).

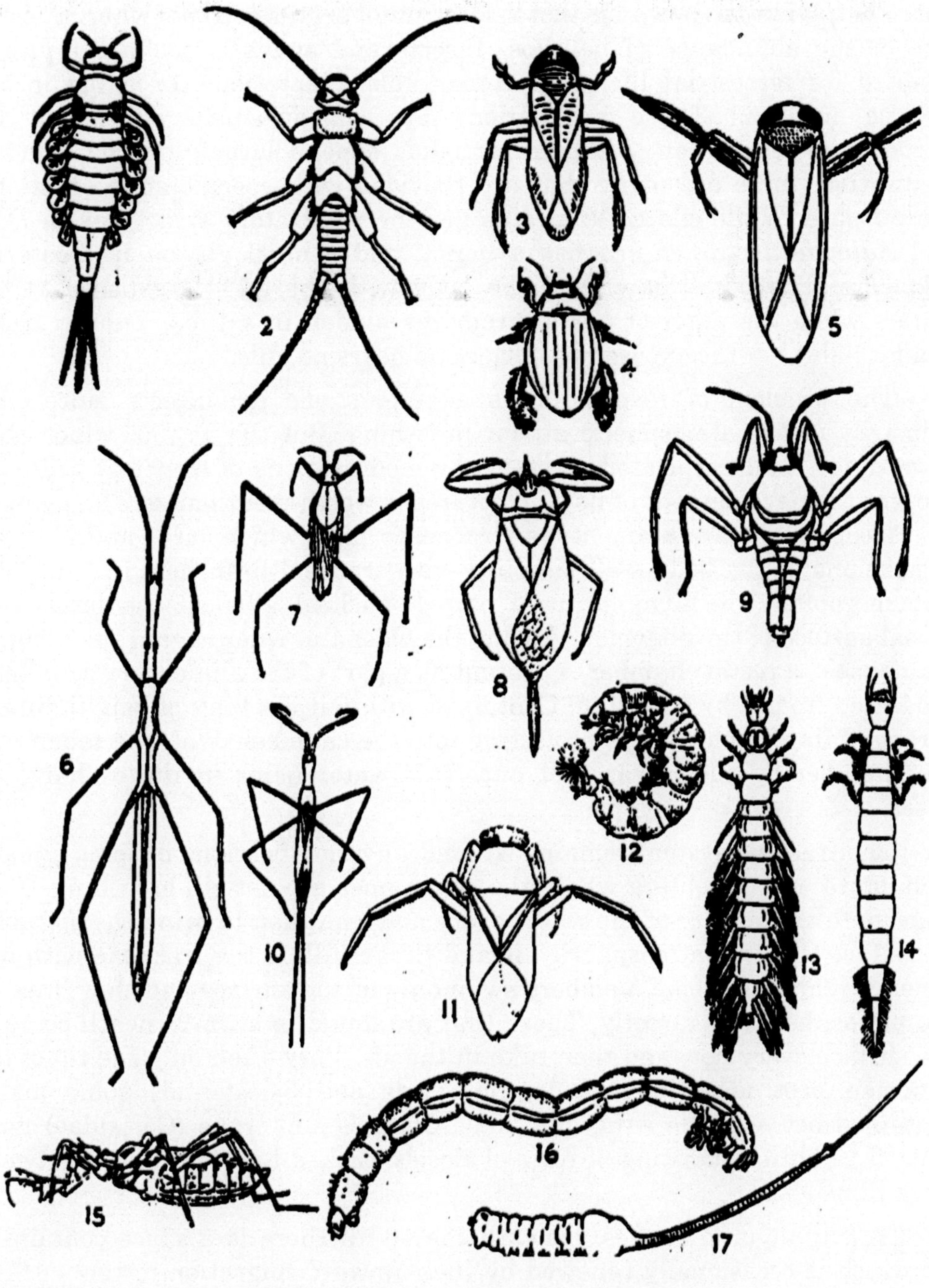

Figure 12.5. Freshwater insect profile

1. Mayfly nymph, 2. Stonefly nymph, 3,5. Corixid bugs, 4. Aquatic beetle, 6. Hydrometra, 7. Gerrid bug (*Gerris*), 8. Water scorpion (*Laccotrephes*), 9. A veliid (water skater), 10. Water scorpion (*Ranatra*), 11. Water boatman (*Notonecta*), 12. Caddisfly larva, 13. A gyrinid larva, 14. Larva of *Cybister*, 15. A naid of dragonfly, 16. A *chironomus* larva, 17. Rat-tailed maggot.

The plankton shows an increase in number and variety when there is rise in the abundance of benthos. Insects and snails which are typically adapted for terrestrial life show remarkable adaptations for a life in the aquatic medium. These adaptations are conspicuously elaborated for purposes of respiration. Although annelids depend largely on the skin for respiration, there are many that are provided with specific adaptations for the purpose. While cilated anal gills are the respiratory structures in *Dero* and *Aulophorus, Brachiura* has a dorsal and ventral gill on its posterior segments. The posterior end of the body with the gills is exposed to the waters while the anterior portion remains hidden in a tube. *Tubifex* has a similar habit but the exposed posterior end bears no gills.

The problem of respiration is a serious one for insects since they primarily utilize atmospheric air for breathing. But the way in which they have tuned their respiratory systems by modifications of structures already present the development of new structures is worth mentioning. The nymphs of Plecoptera bear filamentous tracheal gills which are nothing but protrusions of the body wall supplied with tracheal branches. In the order Ephemeroptera the nymphs have paired tracheal gills on the latral and dorsal surface of the abdomen. Among the Odonata, where again the nymphs are aquatic, a rectal chamber is present as a part of the alimentary canal and separated from it by a septum. Contained within it are rows of longitudinally arranged thin tracheal gills projecting into the chamber. Water is taken into the chamber by suction and let out. This water helps in the exchange of gases.

The tracheal system remains without any modifications in some aquatic Hemiptera and coleoptera where the young ones are not gill-breathing. They migrate to the surface of the water every now and then to bring the spiracles in contact with the atmospheric air and thus refilling the tracheae with air. Among Hemiptera some members swim only at the surface and they breathe the atmospheric air directly. Those that are found to swim beneath come to the surface every now and then take in the air. Very often only the tip of the abdomen protrudes. In corixids, naucorids and belostomids some air is contained between the wing and body. Some hemipterans (Corixidae) have hydrofuge hairs consisting of rows of closely packed hairs that hold between them films of air.

Among aquatic beetles, beneath the elytra there is a space containing air which is occasionally renewed by their upward migration to the surface. The larvae of these beetles which are also aquatic in their habits have either tracheae or tracheal gills. Those with tracheae come up to renew their air supply while those with gills remain permanently under water. Among the Diptera, the adults of which are never aquatic, the immature stages have either rectal gills, or specialized respiratory siphons. But some like the craneflies and hoverflies have well developed trachea which are filled with air by occasional contacts with atmosphere. Larvae of chironomids are capable of anaerobic life. Larvae of Megaloptera have paired cylindrical gill filaments in the initial 7 or 8 segments of the abdomen.

Western Ghats Fresh Water Biodiversity - a profile

The Western Ghats running parallel to the west coast of India from the river Tapti (21° N) to Kanyakumari (8° N) is a prominent landscape feature of the peninsular India. This 1600 km long chain of mountain along with Sri Lanka is also one of the 25 - biodiversity hotspot regions of the world. It is well known for its unusually high diversity of angiosperms and amphibians (Myers *et al.*, 2000). Moreover, the Western Ghats is the catchments of 3 large, 13 medium and 17 minor rivers of the peninsular India (Rao, 1975). These 33 rivers and associated wetlands harbour high diversity of freshwater organisms. A recent study identified the Western Ghats region of India subcontinent as one of the important areas of freshwater fish diversity.

Nearly 750 species of fish are known from the inland waters of India. Out of this 218 with 114 endemics (54%) are known from the Western Ghats (Daniels, 2001 and Bhat, 2002). Species level inventories are incomplete for the freshwater invertebrates of this region. Exception to this is Odonata (dragonflies and damselflies). About 176 species of odonates are known from the Western Ghats, of this 67 (38%) is endemic. More over, most of these endemic donates are restricted to riverine ecosystems. This strongly indicates the diversity of freshwater invertebrate fauna.

Diversity of Aquatic insects of Kaveri River Systems

The aquatic insect fauna of Kaveri river system is remarkably diverse a function of the wide spectrum of microhabitats and trophic resource available, especially in the upper and middle reaches. A total of 112 genera of aquatic insects were recorded in the present study of which Trichoptera (20 genera) and Ephemeroptera (19 genera) dominated the scene. The most diverse trichopteran and ephemeropteran families were Hydropsychidae (7 genera) and Leptophlebiidae (7 genera), respectively in Kaveri river system. Plectopterans are usually well represented.

Longitudinal changes in composition of the Kaveri river communities and in standing crop, generic richness and diversity probably are linked with several physical differences between tributaries and main streams. These include differences in temperature, slope, extent of algal cover and river regulation which invariably modify both flow patterns and discharges. Changes in discharge may affect many other environmental parameters such as current, substrate and water quality. Temperature, a major factor controlling the distribution abundance and life cycles of aquatic insects.

Kaveri Fish Diversity

The Kaveri system in its entirety has 267 species of fishes under 16 orders, 56 families and 118 genera. Jayaram *et al.* (1982) reported 142 species from the Kaveri Basin; they did not cover the estuarine area and also did not conduct an intense survey of some of tributaries. As of now 267 species are known including all earlier reports. Out of these, 103 are restricted to the estuarine stretch. Of these *Salmostoma horai, Mystus keletius* and *Mastacembelus pancalus* are primary fishes which have strayed into the estuarine area. Excluding these three, we have around 100 species which are estuarine or secondary, capable of easy dispersal and are hence excluded from the discussion. However, analyzing the fish fauna as a whole, familywise, we find Cyprinidae as having the maximum (19) followed by Mugilidae and Clupeidae (10 each). As expected, amongst the primary freshwater fishes the carps have always been numerically dominant and they differentiate into many more species than the others.

In most places, species of the small sized cyprinids such as *Puntius arulius, P. sophore, P. ticto,* and *Rasbora caverii* serve as principal forage fishes for the larger predatory fishes as *Aorichthys seenghala, Silondia silondia, Wallago attu* etc. The cyprinids swarm in great abundance in the lowland rivers and highland brooks. Wherever the predator catfish species are absent, numerous cyprinoids themselves evolve into large and *Catla, Cirrhinus, Labeo, Semiplotus, Tor* accompanied the cyprinoids in their geographical radiation, due to the ability of some catfish genera such as *Arius* and *Plotosus* to transgress short stretches of sea. The cyprinoids on the other hand, occupy all available niches in the freshwater zone and have become dominant.

Perciformes occupy the place of Cypriniformes amongst the secondary. The group is represented by 22 families and 74 species (27.72%) in the Kaveri system. As explained earlier, the estuarine zone is the major contributor.

Wetland habitats

Wetlands are lands transitional between terrestrial and aquatic ecosystems where the water table is usually at or near the surface or the land is covered with shallow water. The habitat supports aquatic macrophytes, that grow in water, soil or other substrata that is periodically deficient in oxygen due to water logging. Wetland types are very diverse and it is difficult to demarcate the boundaries. The Ramsar convention of 1971 defined wet lands as "area of marsh, fen, peatland or water, whether natural or artificial, permanent or temporary with water that is static or flowing, fresh, brackish or salt, including areas of marine water, the depth of which at low tide does not exceed six meters". In addition wetlands may incorporate riparian and coastal zones adjacent to the wetlands. As such the comprehensive definition may extend to a wide variety of habitat types including rivers, coastal area and even coral reefs (IUCN, 1989). In 1992, 565 wetlands covering over 6 million ha^2 had been declared as Ramsar sites i.e. internationally important wetlands. In India, chilka lake in Orissa and Keoladeo National park in Rajasthan are Ramsar sites, besides Loktak lake, Manipur, Sambar Lake, Rajasthan, Lakes of Lodakh and wetlands of Kashmir, the brackish water lakes of Rajasthan. Swamps and marshes of North east island chains of Lakshadweep and Andaman, to mention a few. Marshes which are area of shallow water with herbaceous vegetation such as reeds, grasses, sedges are transient ecosystem with fresh water marshes and tidal salt marshes. Being among the world's foremost productive systems marshes have very high productivity and are among the most threatened of wetland types.

Figure 12.6. Wetland birds

The directory of Asian wetlands by IUCN (1989) highlights the predominant wetland types of India's geographic zones. These include:

1. *Tanks, Reservoirs* and the water bodies of the Deccan peninsula;
2. *Backwaters* and *Estuaries* of the west coast of the peninsula;
3. the vast *Saline Expanses* of Rajasthan and Gujarat (mainly Rann of Kutch);
4. *Fresh water lakes* and *Reservoirs* from Gujarat eastwards through Rajasthan and Madya Pradesh;
5. *Deltaic Wetlands* (including Mangroves) *Lagoons* and *salt swamps* of India's east coast;
6. *Marshes*, *Jheels*, *Terai Swamps* and *Chaur* lands of the Indo-Gangetic plain;
7. *Floodplain* of the Brahmaputhra and the *Marshes* and *Swamps* in the hills of northeastern India;

8. *Lakes* and *Rivers* of the montane (primary palaearctic) region of Kashmir and Ladakh;
9. Wetlands (primarily *Mangrove* associations and *Coral reefs*) of India's island areas.

Paddy fields according to IUCN also constitute wetlands which are most extensive, far in excess of all other wetlands put together. They support besides others, considerable population of water fowl, especially egrets, herons, some ducks, storks and cranes (Fig. 12.6). These areas are also being used for aquaculture. Fish such as Catla, rohu, mrigal and tilapia are being cultured in paddy fields.

CHAPTER - 13

ISLAND BIODIVERSITY

The mainland masses of the Oriental and Australian regions are separated from each other by a string of islands, the Malay Archipelago. The large islands of Sumatra, Java and Borneo belong to the Oriental region and their animals are very similar to those of Malaya and in the East, New Guinea and adjacent islands are close to Australia and have many animals in common including marsupials, and cassowaries. In between these two groups of islands are the Philippines; Celebes and Malaccas and Timor Islands that once formed part of the mainland tend to have a considerable part of the fauna of that mainland, for instance, that of Borneo is mainly Oriental and that of New Guinea is Australian. The islands in between have been colonized in a haphazard way by migration across the sea from both the east and west (Fig. 13.1). As such they have individual faunas from those of Malaya and Australia. Celebes (Sulawasi) for e.g. has a hardly any amphibians, no freshwater fish, some peculiar birds and a few mammals. Most of the mammals come from Asia, such as rats and mice, porcupine, civets and deer. Animals peculiar to Celebes are dwarf buffalo and the black tailless baboon (George, 1962).

Besides these groups of islands, the other islands fall into two categories– the *Continental* and *Oceanic islands.* Continental islands are those which at sometime had been part of a large land mass, but which subsequently lost their connection with the mainland and became separated by a stretch of the sea. Great Britain for instance, is separated from Eurasia by the narrow Strait of Dover or as in Formosa, is separated by the mainland of China by a vast sea. Great Britain is a young continental island, but Formosa and Japan are older. Continental islands generally resemble their continents in a general way and have features peculiar to the fauna if they are separated for a longer time.

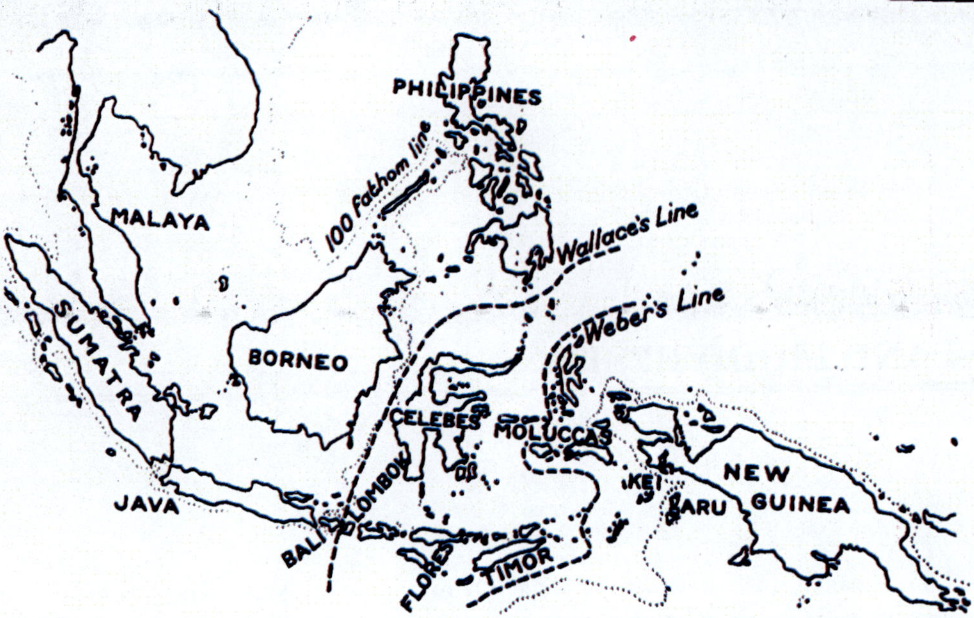

Figure 13.1. Islands in Indian Ocean

The oceanic islands have never had a land connection with a continent and are volcanic in origin and are considerably for off from the nearest land. Of particular interest are the Galapagos islands, the Azores, Bermuda and the Polynesian islands, whose mammals, amphibian and freshwater fishes, will tend to be very rare. Because of the lack of carnivores in oceanic islands certain peculiarities are seen in the evolution of other habitats. For example, birds evolve into ground dwelling forms becoming large and flightless, while others tend to become large like the giant lizards and the giant tortoises of Galapagos island. The evolution of wingless instances is a feature. The Galapagos islands in the Pacific form a group of 15 volcanic islands, about 600 miles from Equador. They have mammals, many land birds and reptiles, but have no amphibians and freshwater fish. A species of bat and rice rats are native and around 26 species of land birds and a penguin species. The famous Darwin's finches are typical, with finches of the same two genera spread all over the islands, but each island fauna is recognizably different from the other. In short, each island had its own species. One of the ground finches showed distinct differences in the properties of the beak and feeding habits. On one island the beak is thin, feeding on cactus; and on another the beak is broader and heavier, the finch feeding on the ground. The Galapagos finches evolved into forms with many different habits, from a seed eating ancestor into cactus feeder and also into wood pecker finches, insectivores, mangrove finches, differing from one another in beak shape and size. Similarly the giant tortoises also varied from island to island. The iguanas which are giant three feet lizards inhabit the land, living on herbivores and feeding on cactus, have become rare, while the marine iguanas are abundant, feeding on seaweeds and land crabs (Fig. 13.2).

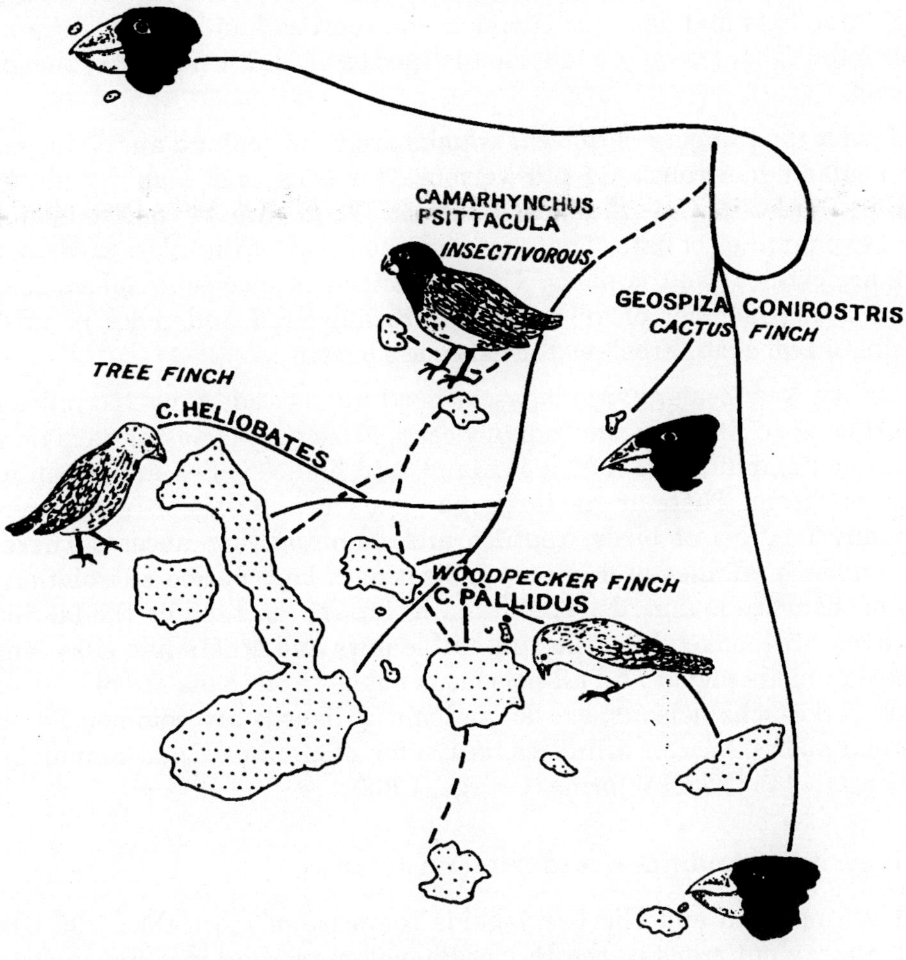

Figure 13.2. Galapagos finches (After George, 1962)

Of particular interest is a tiny group of islands, the Krakatau a small volcanic island in the Sundra strait between south Sumatra and Java, the animal life of which was destroyed by volcanic eruptions in 1883 and repopulation occurred after about 40 years. In 1908, twenty four years after recolonization there were 13 species of birds, (pigeons, kingfishers, orioles etc), 192 species of insects with ants and fishes dominant, 31 species of beetles, six of butterflies. In 1921 during the next survey, there were three times as many species of animals with birds increasing to 25 species, with enormous increase of beetles, butterflies and dragonflies, besides geckos, skinks and python. Twelve years later, in 1933 the fauna further increased, with many species blown to the islands by air currents. Krakatau is a typical oceanic island with no amphibians and no mammals, except bats and rats.

As for continental islands, Great Britain whose modern fauna was acquired mainly from the main European continent, has a large number of sea birds. Shrubs, moles, hedgehogs, rabbits, hares, squirrels, mice, badgers, otters, cats, bats and deer are common and reptiles and amphibia are poorly represented. There is only one endemic species of land vertebrate, the grouse *Lagopus*.

Under the category of *ancient islands* are New Zealand and Madagascar. New Zealand has relicts of old vertebrate stocks and high population of flightless birds, half of which are extinct. There are no native mammals except two families of bats. There is a unique array of flightless birds most of which are extinct like the moas. The characteristic species is *Sphenodon*, the last surviving member of an order that disappeared and there is only one amphibian *Liopelma*. Fresh water fishes are absent.

Unlike New Zealand, Madagascar is a large island, lying 260 miles from the east coast of Africa in the Indian Ocean. Madagascar has more mammals, reptiles and amphibia than New Zealand, and like New Zealand, fresh water fishes are absent. There are no large mammals as in Africa or freshwater fish and many families of birds, reptiles and amphibia are absent. There are many endemic families and genera indicating a long period of isolation and many of them are mammals. The primates are represented by the families of lemurs mainly monkey-like, the ring-tail lemurs and others like silky lemurs. The bird fauna is plenty and all poisonous snakes and toads are absent. Land turtles, geckos, chameleons, skinks and spring lizards are common. In short, the fauna shows African affinities, but some of the birds and amphibia are closely related to Oriental forms (George, 1962).

Biodiversity of Andaman and Nicobar Islands

The Andaman and Nicobar Islands comprise of more than 306 islands and 206 rocky out crops in the Bay of Bengal, spreading in a linear distance of about 1120 km between latitudes 06° and 14° N and longitudes 92° and 94°E. In the entire archipelago, only 38 islands are inhabited, 11 in the Andaman group and the rest in the Nicobar (Fig. 13.3). All these islands are the submarine range of hills that connects Arakkan Yoma of lower Burma with Sumatra of upper Indonesia. The total geographical area is 8,249 km^2 and the islands stretch 700 km from north to south. The fauna and flora of the Andaman and Nicobar Islands are much richer than many other oceanic islands, probably because of their geological history. An extraordinary variety of habitat types, ranging from sandy beaches to coral reefs, mangroves and mountains with dense forests, characterize the Andaman and Nicobar Islands. Some of the large islands such as North, Middle and South Andaman and Great Nicobar islands display a veritable mosaic of forest types. Since most of the islands have very few perennial rivers and streams, the

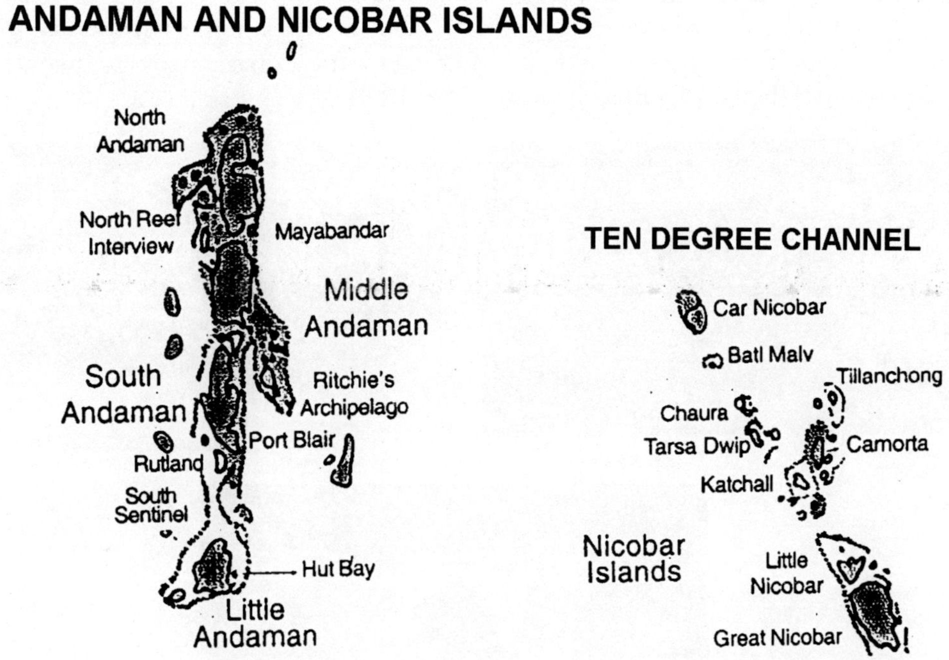

Figure 13.3. Andaman and Nicobar Islands

freshwater wetlands are restricted to only a few major islands. The rainwater fed ponds are valuable sources of freshwater for wildlife and they also serve as a refuge for endangered species, such as the Andaman teal and several endemic amphibians that are habitat specific. The least disturbed and the best-preserved coral reefs and mangroves in India can be found on the Andaman and Nicobar Islands. Their value in conserving soil from erosion and protecting creeks from siltation harbouring a variety of biodiversity cannot be overemphasized particularly in the small islands. The mangrove cover has been estimated to be 929 km^2 in the Andamans and 37 km^2 in the Nicobars. The Andaman and Nicobar Islands have a significant marine ecosystem. Although the land area is only 8,249 km^2, the exclusive economic zone of the island extending to 200 nautical miles from the shoreline is 75 times larger, approximately 6 lakh (0. 6 million) km^2. The astonishing marine biodiversity recorded includes more than 1,200 species of fish, 350 species of echinoderms, 1,000 species of molluscs and many more species of lower forms of life. The Andaman and Nicobar coral reefs are supposed to be the second richest in the world. Coral reefs surround every island and are estimated to cover around 11,939 km^2. They consist mainly of fringing reefs with a barrier reef on the western side. They are the largest reef formations of the Indian subcontinent and contain as many as 177 + species belonging to 58 genera. In these island groups there are two marine national parks viz. Mahatma Gandhi and Rani Jhansi Marine National Parks exclusively recognized for the conservation of coral reefs. Sea grass beds occur in shallow coastal waters and sheltered bays, where clear water allows light penetration. Highly threatened marine animals such as dugongs and marine turtles use this habitat essentially as a feeding ground (Fig. 13.4).

Acropora cytherea, table coral found in the Andaman

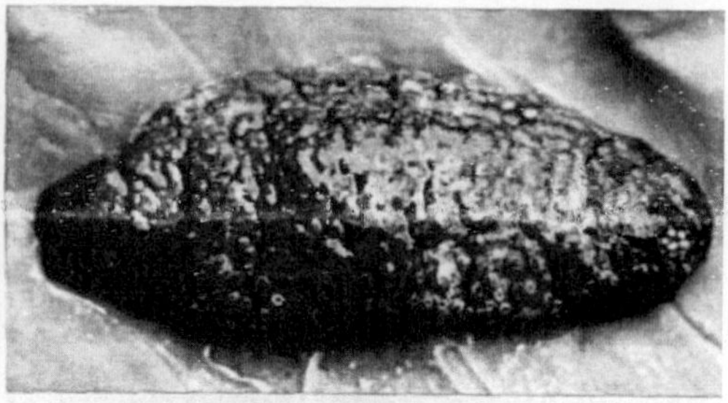

Seacucumber: *Actinopyga mauritiana*

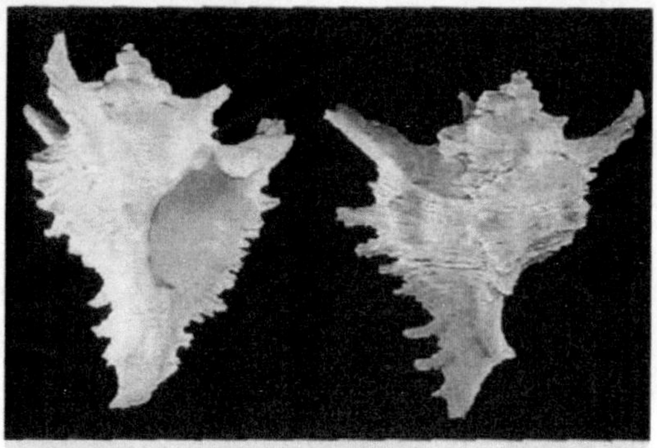

Chicoreus ramosus, ornamental shell from Gulf of Mannar

Figure 13.4. Coral, Sea-cucumber and ornamental molluscs from Andamans (Courtsey, K. Venkatraman)

Biodiversity: The fauna of both the Indo-Chinese and Indo-Malayan regions has influenced the faunal distribution in Andaman and Nicobar Islands. Large mammals are absent in both the Andaman and Nicobar Islands. The geographic isolation of these islands has resulted in a high degree of endemism. The surrounding seas are equally rich in marine biodiversity. However endemism is more pronounced in land animals. The main terrestrial mammals are crab-eating macaque (only in Nicobar), wild boar, civets and several species of bats, rats and shrews. The Andaman horseshoe bat *Rhinolophus congnatus* and Car Nicobar flying fox *Pteropus faunulus*, rats such as *Rattus burrulus*, *Rattus palmarum*, and *Rattus pulliventer* and shrews such as the Andaman Island spiny shrew *Crocidura hispida* and Nicobar tree shrew *Tupaia nicobarica*, are among the endemic mammalian species found here. Endemic subspecies include two subspecies of the wild boar: the Andaman wild pig *Sus scrofa andamanensis* and the

Nicobar wild pig *Sus scrofa nicobarensis*. The other species of large land mammals, spotted deer *Axis axis* and a population of elephants were introduced into the region. Although marine mammals are not endemic, they are often highly threatened. Dugongs *Dugong dugon*, occur in limited and scattered small populations in areas where there are good seagrass beds. The leatherback turtles are reported to be regularly nesting in Great Nicobar Island. The common dolphin *Delphinus delphis* is abundant and is frequently seen in open waters and near the shore. Sightings of blue whale *Balaenoptera musculus* and sperm whale *Physeter catodon* are occasionally reported. With a total number of 270 bird species and subspecies the bird diversity of the Andaman and Nicobar Islands is remarkable. Of these 126 are exclusive to the Andaman and 56 to the Nicobar. Among the well-known flagship species are the nicobar megapode, the Narcondom hornbill and the Nicobar serpent-eagle. The Andaman serpent-eagle, brown coucal and the Andaman tree-pie are some of the other endemic species in the Andaman group. The Nicobar sparrowhawk, Blyth's parakeet and Nicobar bulbul are among the endemics in the Nicobar group. The endemic subspecies include the Andaman flower pecker *Dicaeum concolor virescens*, large Andaman parakeet *Psittacula eupatria magnirostris,* Andaman glossy starling *Aplonis panayensis tytleri,* Nicobar ground thrush *Zoothera citrina albogularis,* and the Andaman teal *Anas gibberifrons albogularis,* to mention some of the more striking ones. The Andaman grey-rumped swiftlet group, whose nests are made entirely of agglutinated saliva, and are of very high commercial value in the international market. Besides, a number of species of snakes, geckos and lizards are also found along with several threatened species like the Andaman water monitor *Varanus salvator andamanesis,* the saltwater crocodile *Crocodylus porosus* and four species of marine turtles, viz, the endangered olive riddley *Lepidochelys olivaea* and green turtle *Chelone mydas*, the critically endangered hawksbill *Eretmochelys imbricata* and the leatherback *Dermochelys coriacea*. All of these species breed in the islands. Of over 1200 species of fish identified on the islands, the deep-sea fish form a commercially valuable resource for the fishery industry. Special mention may be made of the vast diversity of reef fish. The giant robber crab *Birgus latro*, which occurs only in the Southern Nicobar Group, southwestern part of Little Andaman, and south Sentinel Island is a flagship species of the inland invertebrates. There is a remarkable diversity among insects, of which several species are rare and threatened.

Table 1 Faunal Diversity and endemism in Andaman and Nicobar Islands (Courtesy, K. Venkatraman)

Animal Group	No. of species and Subspecies	No. of Endemics	Percentage of Endemism
Terrestrial Fauna			
Mammalia	55	32	61.5
Aves	246	99	40.2
Reptilia	76	24	31.6
Amphibia	18	3	16.7
Mollusca	110	77	70.0
Arachnida	94	38	40.4
Hemiptera	146	22	15.0
Diptera	214	24	11.2
Coleoptera	878	92	10.5
Lepidoptera	426	52	12.2
Isoptera	40	19	47.5
Odonata	36	4	11.1
Annelida	30	9	30.0
Total	**2,366**	**495**	**20.9**
Terrestrial Flora	**2000**	**143**	**14.0**
Marine Fauna			
Mammalia	7	0	0
Reptilia	12	0	0
Pisces	1200	2	0.2
Echinodermata	350	4	0.4
Mollusca	1000+	18	1.9
Crustacea	600	6	1.0
Polychaeta	184	4	2.2
Anthozoa	326+	2	0.6
Porifera	72	--	--
Meiofauna	490	102	21.0
Total	**4241**	**138**	**0.11**
Marine Flora			
Mangrove	**48**	**?**	**?**
Seagrass	**9**	**?**	**?**
Algae	**34**	**?**	**?**

There is also an extraordinary diversity of corals and reef-associated invertebrates. The reefs of rich in soft corals (112 species), and arborescent general like *Acropora, Poecillopora, Seriatopora, Stylopora* etc are abundant, besides the commercially important organisms.

CHAPTER - 14

TAXONOMY AND BIODIVERSITY

It is evident from the forgoing chapters that both terrestrial and aquatic ecosystems harbour countless number of species, not to mention of the vast number of species which have eluded taxonomists or have not been identified due to want of expertise. The marine environment is much more diverse than the terrestrial one in terms of animal phyla, although the terrestrial environment has greater specific diversity. The disciplines of medicine, public health, forestry, agriculture, fisheries, soil science, natural resource management, all depend on studies of biodiversity and in view of human activities accelerating a reduction of this diversity, a better understanding of the systematic study of species and natural groups has become important. Taxonomy involves the recognition and differentiation, identification, description and naming of taxa. Currently more than 1.4 million living species of microorganisms, fungi, plants and animals have been recognized, named and described (Groombridge, 1992; Wilson, 1992). Most of these species are insects (9,50,000) and plants (2,50,000). It is strongly believed that any where between 10 million and 100 million species await discovery (Savage, 1995).

Each species being the product of a long evolutionary history, necessarily stores a vast amount of information in the genome and this information ranges from 1000 genes in a bacterium to more than 4,00,000 in a flowering plant. The amount of DNA stored in the house mouse is 1,00,000 genes, so that biodiversity is a storehouse of information. Species diversity has also a spatial component, the diversity decreasing along several gradients i.e. from low latitudes to high latitudes, from low altitudes to higher altitudes, from warm climates to cooler ones, from terrestrial communities with high rainfall to lower ones. Species richness is at its best in the lowland tropical forests (rainforests) and coral reefs. The impact of human interference has led to species loss in tropical forests in the range of 1-5% per decade, (Savage, 1995; Reid, 1992).

One of the major challenges relates to the continued inventorying of the world's biodiversity. The overriding importance of systematics and its creative role in meeting the emerging needs of science as a whole, have to be recognized, since the relevance of such studies has become increasingly important in the following priority areas (Pushpangadan and Nair, 2001).

(1) Inventorying and monitoring biodiversity (2) biogeography, evolutionary biology and hotspots analyses (3) mapping biodiversity (4) *In situ* and *ex situ* conservation (5) conservation biology of ecologically important and threat prone species and fragile ecosystems (6) community and ecosystem dynamics (7) ecology of indicator taxa, keystone species, flagships and umbrella species (8) chemosystematics (9) molecular systematics involving identification of potential genotypes of economic interest and conservation of genes and nucleotides in DNA libraries (10) phylogenetic systematics (11) environmental impact assessment (12) biomodelling, bioenergetics and bioinformatics. Needless to say that all these categories are dependent or proper inventorying and monitoring of biodiversity, the principal tasks of the Zoological and Botanical Surveys of India, not to mention of other Institutes such as Forest Survey of India, Central Marine Fisheries Research Institute, Indian Agricultural Research Institute, to mention some of the more important. The increased inputs from universities in this direction cannot be overemphasized. The number of species described and estimated living species today are listed indicated below:-

Taxon	Number of Species described	Estimated number of species that exists today
Viruses	5000	Perhapes 500,000
Bacteria	3100	300,000-2.5 million
Cyanobacteria	1700	6000
Algae	40,000	200,000-10 million
Fungi	70,000	1.0-1.5 million
Protozoans	40,000	100,000-200,000
Lichens	17,000	34, 000-56,000
Mosses and liverworts	17,000	35,000
Plants	250,000	300,000-500,000
Animals (Selected groups)	1.1 million	1
Nematodes	15,000	500,000-1 million
Annelids	12,000	50,000
Molluscs	70,000	200,000
Echinoderms	6200	10,000

Crustaceans	40,000	200,000
Mites and spiders	75,000	750,000-1 million
Insects	950,000	8-100 million
Fishes	20,000	40,000
Amphibians	4500	6000
Reptiles	6400	7500
Birds	9100	950,000
Mammals	4000	4100

Some of the major tasks ahead in taxonomic studies should relate to:

1. Inventorying faunistic, floristic and microbial diversity
2. Undertalking revisionary and monographic works on important taxonomic groups.
3. Undertaking conservation biological studies on especially important and threat prone species on priority basis
4. Integration of taxonomic research with ecological, biochemical, molecular and genetic level approaches.
5. Creating databases on various aspects of taxonomy
6. Documentation and maintenance of collections with computerized data accession
7. Documentation of rare, endangered and endemic species so as to aid in conservation

The growing awareness of the benefits and anticipated rapid decline of biodiversity have necessitated a speedy inventory and monitoring of biodiversity at all levels. Multitaxa inventory provides a relatively complete and broader understanding of site, region or global biodiversity (Stork and Samways, 1995; Singh, 2002). Some of the approaches proposed for inventorying the world's biodiversity has been indicated in Table 1 (Singh, 2002).

Table 1. Approaches for biodiversity assessment and inventory (after Singh, 2002)

Level/scale	Purpose
Genetic	Genetic variation among individuals, within populations and among species.
Population	Size of populations for rarity and threat status.
Species	Identity, number, new taxa, geographical distribution, endemism, possible uses.
Single species inventories	High impact species (e.g. keystone, exotic invasive).
	Indicator species (Pollution, disturbance, age indicators, etc.).

	Conservation-focus species (umbrella species, flagship species, threatened species, etc.).
	Species of value to humans.
	Typical or representative species.
Multiple taxa inventories	All taxa biodiversity inventory (ATBI): site-specific, all species
	All biota taxonomic inventory (ATBI): all species of selected taxa global prospective.
	Rapid biodiversity assessment (RBA): focus on certain taxa to provide estimate of biological richness of a site/ region.
Ecosystem and landscape	Type, geographical extent, land cover, unique habitat, threatened ecosystem etc.
Local	Checklists of species, comprehensive inventories.
Regional	Checklists of species can be assembled from local inventories.

Phylogenetic trees are used to represent evolutionary relationship between organisms. The taxonomic units may be species, population, individuals or genes. The present day taxonomic units used to construct a phylogenetic tree are the operational taxonomic units, while ancestral or extinct taxonomic units are inferred from the relationships among operational taxonomic units. In the construction of a phylogenetic tree, nodes and branches comprise the graph. The nodes represent taxonomic units, and the branch lines represent the relationship between them.

Taxonomy and Conservation

Knowledge of individual species is fundamental to maintaining genetic diversity of crops and livestock, developing and implementing biological control of pests, identifying disease causing organisms, discovering medicinal plants, increasing soil fertility. It is in this connection that the discovery of new species has contributed to major economic, health or environmental benefits. Needless to emphasize that the role of systematics is central to the development of long term plans and polices. Taxonomists and ecologists are best equipped to provide the expertise that tends to lead to biologically sound decision. Further, global environmental change is another area where systematics has the potential to constitute the biodiversity. Knowledge of systematics is essential for evaluating the effects of biodiversity that may be produced by changes of global magnitude. It can contribute towards a better understanding of the spread of exotic, invasive and pest and disease causing organisms. Most vector-borne diseases exhibit a distinct seasonal pattern and the significance of the environment in shaping the biology of these diseases is considerable. The entire gamut of the behaviour of arthropod vectors which affect all the links in the chain of disease transmission i.e. the viruses, parasites, vectors, such as mosquitos, ticks, mites, rodents that transmit disease; hosts such as birds which provide a reservoir of disease organisms and predators which exert a control of disease carriers are important considerations (Tyagi, 2004).

As weather changes, habitats change along with the timing of annual processes like flowering, pollination and migration, and species depending on each other do not always adapt to changes at the same time. Freezing temperatures cause lethality to millions of monarch butterflies. At some point, as temperature continues to rise, species will have no more place to run. The natural cycles of interdependent creatures may fall out of synchrony. The polar bear, the pygmy possum, the monarch butterfly are among the threatened species by global warming, not to mention of a host of others, the recognition of which demands a role of taxonomy. The effects of natural disturbances on the local communities of organisms has been documented through bioindicators.

According to Hawksworth and Ritchie (1993), bioindicators are organisms expressing particular symptoms or responses indicative of changes in some environmental influence, usually in a qualitative manner. They are not, however, to be confused with '*Biomonitors*', which are species in which changes in numbers or distribution over time are studied and compared with baseline data. The other ways of using living organisms in monitoring environmental health include *Bioaccumulators* or organisms that accumulate particular substances within their tissues, the concentration of which are measured by chemical methods; *Biomarkers* involving physiological or biochemical changes in an organism caused by chemicals and *Bioprobes* which are symptom-oriented biological effect test system.

Changes in the patterns of distribution of invertebrates, lichens and other organisms (especially amphibians) are often suggested as bioindicators of climate change. Nevertheless, in these and all the above cases, a background knowledge of the ecosystem of interest is of primary importance (Mohan *et al.*, 1995). It can also help in tracking landscape changes through knowledge of species distribution. The need for the establishment of an international network of biodiversity stations cannot be overemphasized.

Alpha taxonomy, the description and classifying of species, is necessary for a better appreciation of ecological questions and spatial organization of genomes, species and communities. Taxonomic information is essential for addressing many conservational issues, in particular the spread of alien species, conservation of migratory birds, emergence of new diseases and the like. Diamond (1987) is of view that "all decisions about conservation are based on faunal and floral catalogues defined by the information that specimens provide about species and races, their geographical variation and distribution". Van Wright (1992) proposes a biodiversity conservation strategy based on systematic analyses of geographic ranges and taxonomic relationships of a wide variety of taxa, leading to the identification of effective global, natural and local networks of protected areas". Thus taxonomy contributes as a foundation for the very future of species and promoting increased taxonomic knowledge is a worthy goal to advance our understanding of biodiversity. Properly assessed and well interpreted museum collections can provide important baseline information for designing land-use and agricultural pest management programmes. The ideal is to establish a *'biodiversity commons'* where biodiversity information should be free and openly accessible and this alone will raise the profile of taxonomy and increase the number of people who actually use the fruits of taxonomic research (McNeely, 2002); Moritz, 2002; Godfray, 2002). In this connection mention may be made of the role of Bionet-International which is a global network for biosystematics of arthropods which offers assistance in training taxonomists, as well as computer aided taxonomic tools enabling wider use of taxonomy. For biodiversity harvesting and bioprospecting, inventorying has a major role to play; since it is very essential to identify the organisms at species or subspecies levels, which is also necessary for technological uses of biodiversity (Gauld, 1999).

Systematics and Global Change

Global environment change is another area where systematics has the potential to contribute to biodiversity studies. The replacement of complex natural ecosystems by monoculture agriculture, besides increased industrialization and urbanization have led to the conservation of additional natural habitats into marginally productive agriculture use. Concurrently mechanized land clearing and agriculture, industrialization and urbanization have added a variety of trace gases and pollutants to the atmosphere. Deforestation and atmospheric contamination together with diversity of land and water use pattern have combined to threaten global changes including major climate warming, modified rainfall patterns, rising sea levels, enhanced UV radiation, depletion of soil fertility and decreased water availability, land use change, loss of biodiversity, invasion of exotic and pest species and pollution-induced changes in atmospheric and water quality. All these have influenced the biotic composition of ecosystems.

Systematic knowledge is essential for elevating the effect on biodiversity that may be produced by changes in global magnitude. Systematics can contribute to understanding and tracking climate and other global changes in several ways, such as (a) providing baseline data for monitoring changes in biodiversity (b) identifying changes (c) tracking landscape changes thorough knowledge of species distribution (d) identifying spread of exotic invasive and pest or disease causing organisms. It is here that there is a need to have an international network of biodiversity stations to provide the data. The need to discover, describe and inventories biodiversity at various sites becomes important.

Systematics is the fundamental science of biodiversity, basic to all aspects of biology, with the principal mission relating to the discovery, description and naming of species and higher taxa with a knowledge of their phylogeny. Its role in providing essential environmental services and usable resources, which are data necessary to preserve and use biodiversity, is therefore significant. Systematics is vital in assessing the discovery of new and improved food crops, medicines, besides determining the know-how to preserve certain areas to maintain biodiversity. In short, systematics has a central role to play in biodiversity studies, since it can "provide base line data and monitoring to track global climate and ecosystem change, including rates of speciation decline, expansions and extinction, besides ecosystem modification or degradation and the spread of exotic pests or disease vectors" (Savage, 1995).

To quote E.O. Wilson (1985) "If systematics is an indispensable handmaiden of other branches of research, it is also a fountainhead of discoveries and new ideas, providing a remedy for what biologist and philosopher William Morton Wheeler once called the "dry rot" of academic biology. Systematics has never been given enough credit for this second, vital role. If a biologist can identify only a limited number of species, he is likely to gravitate toward them and end up on well-trodden ground; the remainder of the species remain a confusing jumble. But if he is well-trained in the classification of the organisms encountered, his opportunities multiply. The known facts of natural history become an open book, patterns of adaptation fall into place, and previously unknown phenomena offer themselves conspicuously. By proceeding in this opportunistic fashion, the biologist may discover a new form of animal communication, a previously unsuspected mode of root symbiosis, or a relation between certain species that permits a definitive test of competition theory. The irony of the situation is that such successful research then gets labeled as ecology, physiology, or almost anything else".

Systematics is linked not only to the remainder of biology, but to the fortunes of the international conservation movement, which is now focusing its attention on the threatened environments of the tropics.

CHAPTER - 15

BIODIVERSITY AND BIOTECHNOLOGY

The increasing application of biotechnology and molecular tools to biodiversity studies has led to the augmentation and value of bioresources. This is well evident in the increased productivity in the food front, better human health and increased environment friendly technologies. Biotechnology or the use of technologies based on living system to develop commercial processes and products, include such techniques as recombinant DNA, gene transfer, embryo manipulation and transfer, plant regeneration, cell culture, monoclonal antibodies and bioprocess engineering. It has been possible to genetically alter certain crops to increase their tolerance to certain herbicides. Biotechnology has also been used to develop safer vaccines against viral and bacterial diseases. In short, biotechnology offers tools for a better understanding of living systems, of our environment and our own selves. In the field of agriculture, biotechnology offers tremendous potential for increasing crop and animal production and bio-processing. It enables new approaches to develop high yielding and more nutritious crop varieties, improved resistance to disease, reducing the need for fertilizers. Bioprocessing which is the use of living systems or their components to create new useful products such as food products, treatment and use of wastes and using renewable resources as fuel. Besides biotechnology can also improve forestry and it products, fiber crops and the like. In order to realize the benefits, it is necessary to resort to research in gene identification, gene regulation, structure and function of gene products, cellular techniques, besides understanding the behaviour and effect of genetically engineered organisms in the environment. In terms of biodiversity, it provides tools for a critical assessment of biodiversity in terms of genetic diversity of enabling better use of bioresources.

Biotechnology and Taxonomy

Biodiversity has come to assessed more on genetic basis and not solely on the basic phenotypic traits, since DNA and RNA sequences provide a better basis for assessment. In this connection the increasing role of biotechnology in biotype studies in insects has come to be recognized. Biotypes are morphologically similar, but functionally different, and in today's scenario enough information has been forthcoming on the genetic basis of interaction between biotypes and host genotypes in support for gene-for-gene interactions. DNA associated molecular marker technology is of use in breeding for desirable resistance, with a diverse array of markers. Plant-insect interaction and biotype specific resistance are better understood by the use of molecular and genomic techniques. Taxonomic system based on DNA and RNA data or on the basis of variation of gene products could play an important role in the assessment of biodiversity.

Genetic engineering in agriculture has come to be recognized as a method to improve yields and reproduction, disease resistance, better quality products enabling transformation dramatically by increasing efficiency and productivity. Some of the areas of research now in vogue are locating and identifying agriculturally important genes and creating chromosome maps; understanding the mechanism of regulation and expression of the genes; understanding the structure and function of gene products in metabolism; developing refined techniques for cell culture, cell fusion, regeneration of plants; understanding the complex physiological and genetic interactions, besides the behaviour and effect of genetically engineered organisms in the environments (NRC, 1987). In short, the tools of biotechnology, combined with traditional techniques in biology and chemistry increase enormously both the power and the pace of research in agriculture systems. The tools of biotechnology can provide researchers with new approaches to develop higher yielding and more nutritious crop varieties, to improve resistance to disease and adverse conditions and reduce the need for fertilization and other chemicals.

Another area in which genetic engineering technology works prove useful is in developing biological pest control methods. The potential exists to identify genes controlling the properties and transfer of these traits to plants. Extensive basic research on biosynthetic pathways of hormones has led to their utility value in altering crops so that they produce their own repellents. A notable example is the role of baculoviruses as carriers of neurohormonal genes. One of the major successes relates to the genetically altered bacterium *Bacillus thuringiensis* which produces endotoxins. Other potential insect resistance genes coding for inhibitors of digestive enzymes such as proteinase and amylase inhibitors and lectins have been discovered and are in operation and new plant-transgene combinations are being developed. Transgene expression has been augmented today by the use of green house-specific and pollen-specific promoters, with ranging expression levels in different tissues.

Manipulation of the genomes of crop plants and insect natural enemies has opened up new avenues for biological control, while transgenic plants may increase biological control efficiency, the development of transgenic natural enemies that have been modified to tolerate insecticides and adverse environmental conditions, has augmented biological control applications. Of particular interest is the utilization of a predatory phytoseid mite *Metaseiulus occidentalis* in biological control through genetically enhancing natural enemy efficiency for integration into pest management programmes.

Critical analyses of several organic compounds have become possible by utilizing biotechnological tools. Quick screening of thousands of plants and microbes has become possible through development of very sensitive bioassays to identify those with potential in agrochemical, pharmaceutical, food and other industries. DNA sequences of several organisms have been unraveled enabling identification of specific gene products, besides enabling identification of desirable genes present in the wild relatives of plants and microbes. This has resulted in the transfer of such genes to domesticated plants to produce transgenics. Further, transgenic organisms can be used as sources of proteins and peptides, lipids and fatty acids, carbohydrates and secondary metabolites which are useful in a host of processes relating to flavours and aromas, food additives, pharmaceuticals and even biopesticides. The improvement, maintenance and rehabilitation of our environment is possible through identification of soil microbes through DNA finger printing, through engineering and transferring of important microbial genes in enriching and cleaning up of degraded soils. Such bioremediation has become possible in a variety of environmental problems relating to a host of pollutants. Depletion of biodiversity through man-made environmental degradation could be restored or overcome by the use of biotechnology, wherein engineering and transferring of key microbial genes involved in the enrichment or cleaning up of water or soil have become key process, for the restoration and remediation of degraded and contaminated ecosystems. Bioremediation can be carried out by suitable microbial system *in situ*, the microbes multiplying at the expense of pollutants thereby enhancing remediation.

Molecular biotechnology techniques are therefore available for transferring resistant genes from other sources to the crops. Related organisms can be modified to compete with the pathogen or pest and modify their virulence. Prospects of introduction of genes into parasites that could reduce their virulence are on the cards. This is certainly a positive approach, since the development of molecular biology techniques to obtain and handle DNA has provided the tools for transferring and evaluating genetic characters from different organisms.

CHAPTER - 16

REMOTE SENSING AND BIODIVERSITY

Satellite based remote sensing enables comprising details on the surface of the earth covering large areas, besides facilitating more frequent observations by periodic reappearance at the same place. Two technologies, remote sensing and geographic information system (GIS) have opened the door towards a better understanding of the distribution and status of biodiversity over large areas and lending themselves to mapping biological resources. There are three basic aspects in this analysis – vegetation, species distribution and the location of protected areas. GIS comprises a powerful set of tools for collecting, storing and retrieving, transforming and displaying special data. Computers store information and different set of data, such as soil type, vegetation cover, gradient etc can be overlaid to give a comprehensive description of a mapped area. GIS provides information on specific location and allows quantification of particular spots, helping local planners for assigning priority to future land conservation. Remote sensing technology uses the visible, infrared and microwave regions of the solar radiation to collect information about various objects of the earth's surface. The different regions and corresponding wavelengths in the case of electromagnetic spectrum are as in Figure 16.1. Different objects manifest themselves predominantly in certain wavelengths of the electromagnetic spectrum in the visible range with certain specific wavelength ranges (Narayanan, 1999). Within the visible region rainbow colours are utilized, but when these colours merge, we see it as white light. The use of filters and sensors active in certain regions of the spectral range of colour give us information in different bands of the electromagnetic spectrum. The spectral range and application areas are as under:-

Band	Spectral Range	Application areas
1.	0.45-0.52	Coastal environment studies (costal morphology sedimentation studies), soil/vegetation differentiation, coniferous/ deciduous vegetation discrimination
2.	0.52-0.59	Vegetation vigour, rock/soil discrimination, turbidity and bathymetry in shallow waters
3.	0.62-0.68	Strong chlorophyll absorption leading to discrimination of plant species
4.	0.77-0.86	Delineation of water features. Land forms/geomorphic studies

However limits do exist, as for instance, when there is cloud cover, no data is obtained. All the same, by using sensors giving data at the microwave region of the EM spectrum, these limitations are overcome. The first Indian Remote sensing satellite (IRS – 1A) was launched on March 17, 1988 with 22 day repeat cycle and orbiting at an altitude of 904 km (Narayanan, 1999). Three types of resolution in remote sensing from satellite exist: spectral resolution, spatial resolution and temporal resolution.

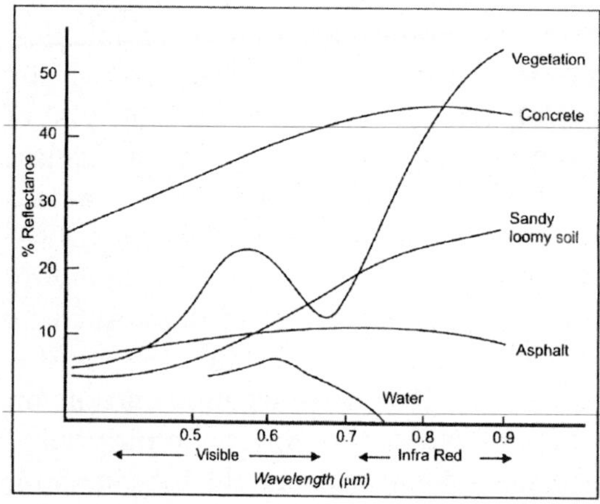

GAMMA RAYS	0.003 A 0.01 μm
X-RAYS	0.1 μm
ULTRAVIOLET REGION	0.4 μm
VISIBLE REGION	0.7 μm
NEAR INFRARED REGION	1.5 μm
MIDDLE AND FAR INFRARED REGION	1.0 μm
MICROWAVES	0.8 metres
RADIO WAVES	10 Kilometres

(μm is micrometer which is 10^{-6} of a metre)
(A – Angstrom – or 10^{-10} metre)

Figure 16.1. Spectral reflection curves (left); different regions and wavelengths of EMS (right) (after Narayanan, 1999)

Mapping of coral zones provides information about how to conserve their ecological balance, save the mangroves and implement effective coastal zone management. The satellite remote sensing enables us to locate and monitor tidal wetlands, coastal landforms, potential agricultural sites, mangroves, estuaries, shore line changes, potential aquacultural sites besides monitoring pollution one to oil slicks. Mangroves form an important vegetation belt along the coastal region having considerable influence on coastal ecosystems. Remote sensing data not only help monitoring the whole coast, but can also select suitable locations for prawn/shrimps farming. The National Remote Sensing Agency (NRSA) was established in 1975 in Hyderabad to develop and utilize operational capability of remote sensed data for survey and management of national resources of the country. The first operational Indian Remote sensing satellite IRS-IA was launched in 1988.

Remote sensing is a major tool to collect data relating to monitoring changes in forest cover, assessing land use and forest degradation, evaluating the productivity land and providing inputs for forest development, management and strategic planning. The study of spatial distinction of forest types is to provide better insight of the types of species to be introduced either by replacement of regeneration of degraded forests in a more useful way. Preparation of forest type maps is now possible using satellite sensed data. Besides mapping forest types, density and stock mapping, monitoring deforestation and afforestation, monitoring enhancement in forest areas, assessment of production potentials and wild life studies using remote sensing data from different satellite like lanscet, IRS 1A and 1B are now possible. Also the identification of forest fire damaged areas, periodic and seasonal data of specific forest areas prone to fire could be used by forest managers to enable taking fire prevention measures. Effective forest management could be had by using GIS techniques by integration informations obtained from diverse sources facilitating organization of a data base, data updating and retrieval of relevant data even at the district level (Narayanan, 1999). Conservation and restoration processes especially with regard to renewable natural resources like forests are essential to contribute towards environmental stability. For this, remote sensing and GIS are reliable modern techniques (Fig. 16.2).

Multipurpose river valley projects such as Idukki in Kerala, Periyar river basin, and other water resource projects in terms of water logging, salinity caused poor drainage system, sedimentation of resources, submergence of forest land, change of land use patterns and related environmental impact have been studied by remote sensing as a tool. This has enabled preparations of maps pertaining to forest/vegetation cover, soil classes and the like in river basins. The use of remote sensing and GIS on coastal systems in relation to brackish water aquaculture has proved by useful in producing thematic maps involving soil types and water quality.

Wastelands can be mapped through remote sensing and make efforts to bring more land area with use for productive purposes. Besides satellite data has a direct bearing on forest management, in various applications like inventorying, checking forest fires, biomass and fuel-wood assessment. More detailed aspects relate to dynamic changes of forest cover and forest type distribution. Plantation monitoring, forest inventory, preparation of zones, assessment of ecological and biodiversity related measures, forest change and conversion, forest damage due to fire, grassland assessments and the like are not only possible, but are also in vogue.

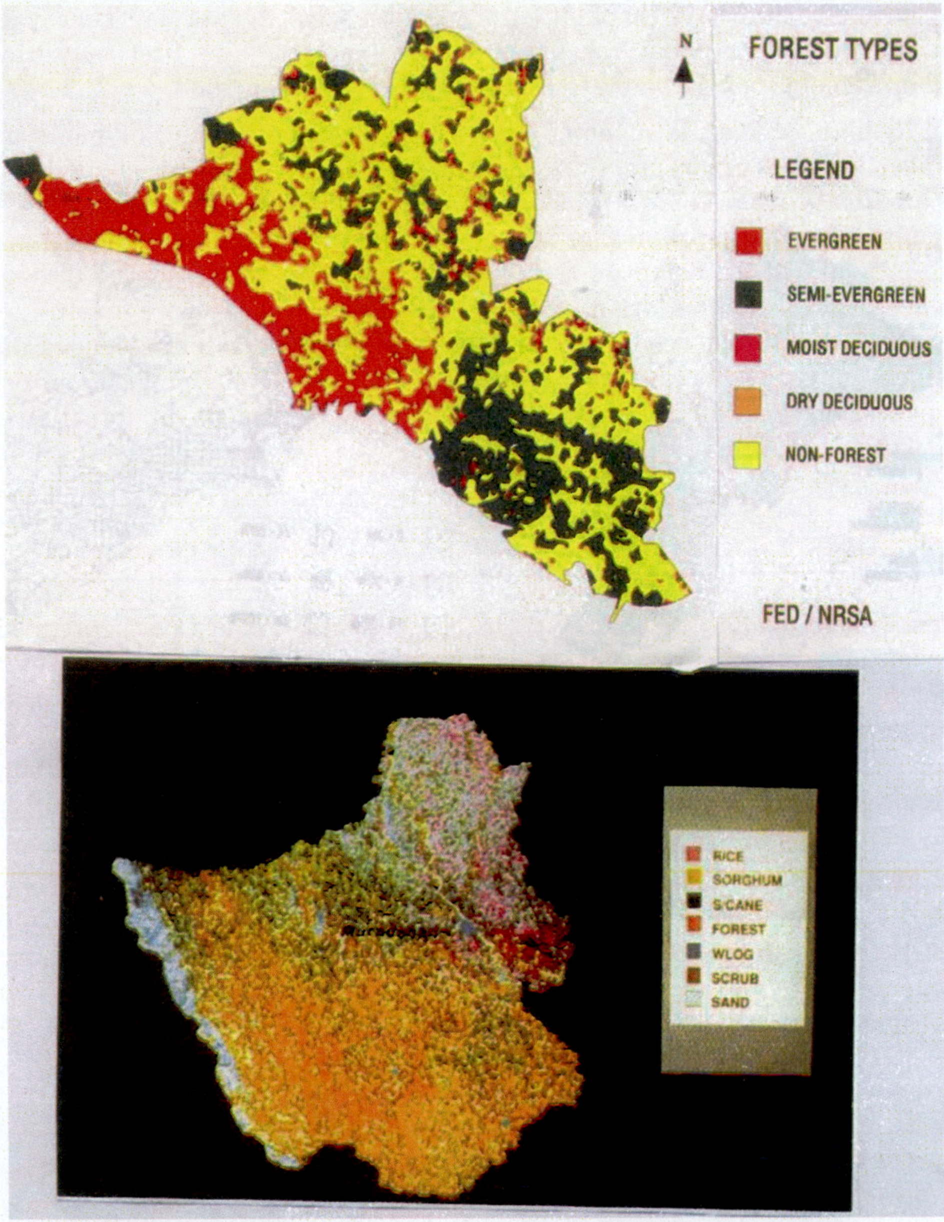

Figure 16.2. Forest types (Top) and Agriculture area (bottom) as shown by remote sensing (Courtesy, Current Science).

The satellite remote sensing techniques are also used to generate reliable data on existing irrigation schemes, and helps better water management, besides sustainable agricultural productivity, cropping pattern, soils, water resources and climate. Satellites have improved their capacity to monitor and the launching of useful satellites with high quality GIS for more detailed microlevel planning programmes, efficient and suitable management of our natural resources have been made possible.

Remote sensing offers excellent possibilities to map and monitor urban sprawl, particularly when cities are becoming fast overcrowded necessitating a planned and regulated urban growth. For this purpose the importance of various types of remote sensing data for urban planning are available such as physical features, involving topography, soil associations, geological base and minerals, drainage pattern, vegetation, water bodies, climate etc. Besides the nature of agriculture, industry, forestry as well as communication network, land use patterns, housing density and so on become important.

One of the instances of great relevance relates to the use of remote sensing in disaster situations such as the recent Tsunami incident. ISRO has supplemented satellite images by deploying an aircraft equipped with high resolution imaging system. The aircraft has been used to obtain pictures of the Tsunami affected areas in Tamil Nadu and in the Andaman and Nicobar islands.

CHAPTER - 17

CONSERVATION AND MANAGEMENT OF BIODIVERSITY

The scientific world became conscious of the impending 'biological diversity crisis' during the early 1970s, but a common forum to address this issue was non existant at that time. Conservation biologists the world over keenly felt the urgent need to formulate strategies for conserving our precious biological diversity, including the range of species, genetic variation, biological communities and ecosystem interactions. The need to immediately check the rapid extinction of species due to varied reckless anthropogenic activities, besides maintaining the intricate interactions of species in natural communities become very relevant, in terms of the inherent value of biodiversity. At this stage, a subtle distinction between the terms 'preservation' and 'conservation' became obvious. 'Preservation' implies the protection of biodiversity from any kind of human activity or interference, while 'conservation' implies the protection of biodiversity for sustainable utilization (Virchow, 1998). Thus, the science of conservation biology was known as a 'crisis discipline' when the ecologist Michael Soute organized the first international conference on conservation Biology in 1978, which met at the San Diego wild life animal park, providing scope for wild life conservationists, zoo managers and hardcore biologists to discuss their common interests. This discipline took its final shape in 1985, when the theoretical aspects of population biology was combined with the practical experience of wildlife, forestry and fishery management to evolve meaningful conservation modalities. The conservation of biodiversity is a holistic concept and encompasses a whole spectrum of biota and of activities ranging from ecosystems at the macro level (*in situ*) to DNA libraries at the molecular level (*ex situ*) and everything in between.

CONSERVATION STRATEGIES

1. Conservation of diversity at the level of populations and genes

It is the usual practice to create national parks and sanctuaries for protection of "charismatic" big animals like lions, tigers and rhinos. However, protecting the few remaining population of especially endangered and rare species is the key to preserving species. To ensure the long term survival of a species, a minimum variable population (MVP) has to be maintained. MVP is the smallest population size that is likely to have a very high chance of persisting for the foreseeable future. The next step is to study the home-range size of individuals and colonies of an endangered species, its minimum dynamic area (MDA), the area of suitable habitat necessary for maintaining the minimum viable population should be estimated. These steps are vital for conservation of endangered species consisting of a just a few populations, or even a single population, because small populations are subject to a more rapid rate of extinction for three main reasons viz. loss of genetic variability, inbreeding depression, demographic fluctuations and environmental variations or natural calamities. Once a population size has been reduced by habitat destruction, fragmentation and other human activities, it is even more vulnerable to random fluctuations in size, leading to rapid extinction which will increase the drive to further extinction. It is obvious that a population is the most reasonable level at which genetic conservation is to be attempted. For example, electrophoretic studies of protein (LDH enzyme) on patterns of the one-horned Indian rhinoceros (*Rhinoceros unicronis*) in Chitwar National Park in Nepal, indicate that just a small total population of around 1500 animals have remarkable genetic diversity. This is maintained despite passing through a population bottleneck, because they have successfully migrated into the park from a large area, thus ensuring long-term survival. However, the deadly threat of habitat loss is looming large for the Indian rhinos. The geographic range of this animal, originally encompassing northern India, Pakistan. Nepal, Myanmar and Bangladesh has considerably shrunk due to human settlement. Indian rhino populations today are confined to the Kaziranga sanctuary in Assam. Thus, population is the most appropriate level at which genetic conservation can be attempted and should be followed wherever possible, once the ecosystem and species conservation has been achieved satisfactorily.

Protection of endangered species involves establishing new wild populations of those species. New populations of rare and endangered species can be established in the wild using either captive-raised or wild caught animals. Reintroduction involves releasing individuals within the historical range of the species; introduction involves release of individuals at a site outside the historical range of the species; augmentation involves releasing individuals into an existing population to increase population size and genetic variability.

2. Conservation of Species Diversity

In the conservation scenario as a whole, 'species' as an entity plays a crucial role conceptually, biologically and legally. This is amply testified by the importance given to species, while developed and developing countries enact legislations on conservation. An instance in point is CITES (Convention of International Trade in Endangered species of Wild Fauna and Flora). An understanding of the biology, species-area relationships and life history patterns of constituent species is very essential even for conservation from the point of view of habitats and ecosystems and for the design of management of bioreserves. While prioritizing species for conservation, threatened species receive topmost attention.

The major IUCN threatened categories (IUCN Red List Categories, 1995) currently recognized are as follows:

1. *Extinct* (*Ex*): species that have been totally lost during the past 50 years. Species lost in the wild while living under cultivation is categorized as 'Extinct in the wild' or EW.
2. *Endangered* (*E*): Taxa in danger of extinction and whose survival is unlikely if the causal factors continue operating. Included are taxa whose numbers have been reduced to a critical level or whose 'habitats' have been so drastically reduced or modified that they are in immediate danger of extinction. Also included are taxa that are possibly already extinct, but have definitely been seen in the wild in the past 50 years.
3. *Vulnerable* (*V*): Taxa believed likely to move into the 'Endangered' category in the near future. Included are taxa of which most or all populations are decreasing or being lost to disturbance and fragmentation; and taxa whose biogeographical range is decrea-sing through loss of peripheral populations.
4. *Rare* (*R*): Taxa with small or limited geographical range popula-tions that are not at present 'Endangered' or Vulnerable', but are at risk.
5. *Interminate* (*I*): Taxa known to be 'Endangered 'Vulnerable', or Rare' but where there is not enough information to categorize.

Unknown Categories

i. *Insufficiently known* (*K*): Taxa that are suspected but not definitely known to belong to any of the above categories, due to lack of information.

ii. *Status Unknown* (*?*): No information

iii. *Candidate* (*i*): Taxa whose status is being assessed and which are suspected, but not yet definitely known to belong to any of the above category.

Not evaluated category

A taxon that has not been assessed against the criteria comes under this category.

The first category of species to be given priority are the Threatened species, exhibiting rarity. Rabinowitz, Cairns and Dillin (1986) have examined types of rarity and distinguish 3 traits characteristic of all species.

1. *Geographical range*: Whether a species occurs over a broad area or whether it is endemic to a particular small area.
2. *Biotype specificity*: The degree of which a species occurs in a variety of biotopes (habitat) on is restricted to one or a few specialized sites.

A species selected for conservation is routinely assessed for a range of factors leading to rarity. The preference for endemic taxa (taxa restricted in distribution to a specified region) is one possible and positive approach. It is also advisable to do *cladistic prioritization*, which is a method to assess and to some extent quantify the distinctiveness between taxa considered for conservation. It is based on the phylogenetic relationship between species arrived at through cladistics. The following categories of species are invariably given priority in species centered conservation programmes.

i. *Indicator species*: This may be a taxonomic indicator, or an indicator of global warming or habitat health or bioindicator sensitive to pollutants, human interference, ecological instability on other disturbances.

ii. *Umbrella species*: that usually require extensive area providing protection for other species within the ecosystem.

iii. *Keystone species*: that are important to maintain the ecological integrity of the community and are essential to survival of other species.

iv. *Charismatic species*: which are popular for collection and culture and provide aesthetic value.

v. *Recreational species*: which are popular for collection and culture and provide aesthetic value.

Species-based approach has resulted in identification, at the national an international levels of taxa that need conservation and which are then listed by IHCN or national agencies in Red Data Books or detailed in the IUCN species survival commission (SSC).

3. Conservation of Ecosystem Diversity: Proponents of holistic approach to conservation recommend this strategy to ensure that representative areas of ecosystems or important habitat sites are maintained through a network of protected areas. This approach is especially suitable for conserving tropical rain forest ecosystems from immediate collapse even prior to collecting detailed information regarding the status and distribution of its constituent species. An innovative approach based on ecosystems to conservation planning called *Conservation potential/Thread Index*, which forecasts how current deforestation rates would affect the conservation and establishment of protected areas (Dinerstein and Wikramanayake, 1993).

An ecobiogeographic approach integrating both species and ecosystem-based approaches is strong by advocated bearing in mind the importance of community-based conservation measures with active involvement of local people and tribals who actually own biodiversity. Before it is too late, areas or ecosystems of high diversity, especially the threatened species are to be identified and concrete steps are to be taken conserve such areas.

With a view to effectively combine both species and ecosystem based approaches, to conservation, Ganeshaiah *et al.* (2001) proposed two complementary approaches viz. *Top-down* and *Bottom-up protocols* geared to address the immediate conservation needs and the long-term needs to conserve the whole genetic resources of an ecosystem. In the top-down approach, the major landscape elements endowed with a diversity of genetic pools of known and unknown species are first identified. Then the large landscape patches with greatest genetic diversity of a set of species and with a high level of biodiversity are identified for conservation actions. In the Bottom-up approach, pyramiding the species that need conservation attention on priority basis is done. Judicious admixture of these two approaches is efficient and cost-effective in conserving species in the tropics.

In situ and *ex situ* conservation (Fig. 17.1)

The different systems of conservation in vogue today are *in situ* conservation, *ex situ* conservation besides *in situ* on-farm conservation of vital importance to global food and health security (Swaminathan, 2002).

Ex situ Conservation

Conservation in *ex situ* gene banks ensures that stored material is readily accessible; can be well documented, characterized, and evaluated; and is relatively safe from external threats. When material is stored in this way, plant evolution is effectively frozen at the time of storage.

Of the main *ex situ* methods of conservation, the most common is the storage of dried seeds in gene banks at low temperatures. For recalcitrant seeds, such as those of many tropical perennial species, and vegetatively propagated germplasm, such as *Musa*, cassava, or potatoes, other methods are needed. These include conservation as living collections in field gene banks or *in vitro* either as living plantlets, as plant tissue on appropriate media, often under conditions of slow growth, or by cryopreservation at very low temperatures, generally using liquid nitrogen. Genetic resources are also conserved as frozen or freeze-dried pollen. Increasing use is being made of isolated genomic DNA banks for the storage of germplasm (Adams, 1997) and for most major crop species as cDNA inserts in microbial species such as yeast and bacteria. DNA nucleotide sequence data for functional genes are a relatively new mode of conservation of genetic resources. From these data, polynucleotides can be synthesized in the laboratory.

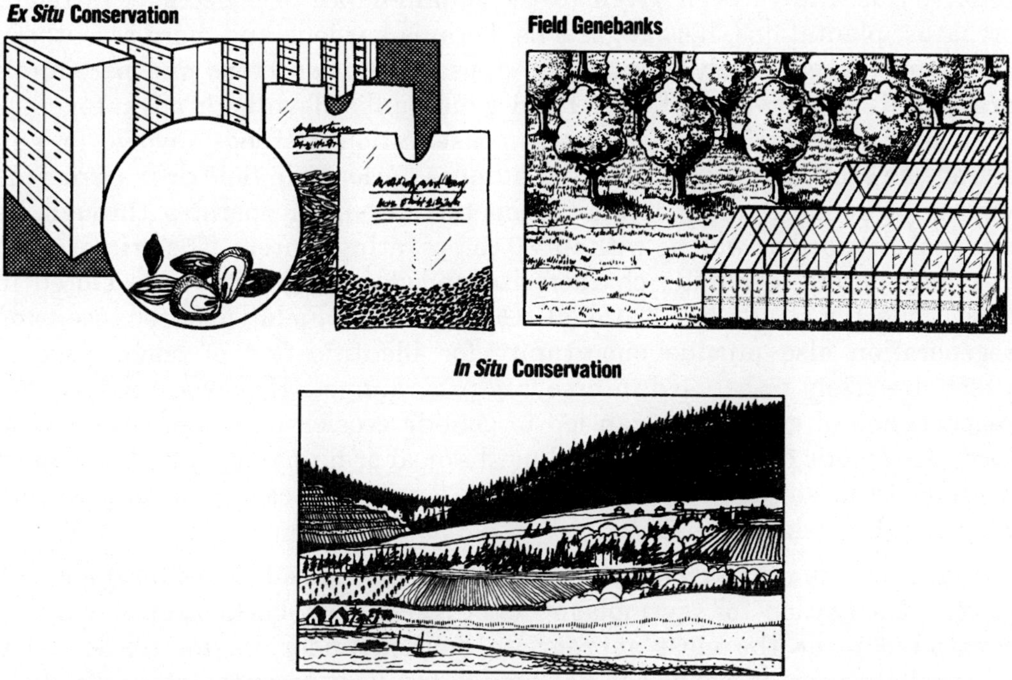

Figure 17.1. Diagrammatic representation of *in situ* and *ex situ* conservation (after WRI, IUCN, UNEP, 1992)

In situ techniques allow the conservation of greater inter- and intraspecific genetic diversity than is possible in *ex situ* facilities. They also permit continued evolution and adaptation to take place, whether in the wild or on-form where human selection also plays a critical role. For some species, such as many tropical trees, it is the only feasible method of conservation. Sustaining habitats indefinitely due to hazards such as extreme weather conditions, pests, and diseases is a major concern for in situ conservation. Difficulties in mapping, characterizing, evaluating, and accessing genetic resources *in situ* are evident.

As with *ex situ* conservation, the method adopted depends on the nature of the species. Traditional crop cultivars may be conserved on-farm, while undomesticated relatives of food crops may require the setting aside of reserves. Agroforestry species, and other plants which require little maintenance, can be conserved by developing and maintaining sustainable harvesting practices and involving local communities, while forest genetic resources are usually maintained in forest reserves and in areas under specially designed management regimes.

Field gene banks and clonal repositories

Under *ex situ* methods employed for germplasm conservation, high priority has always been given to the maintenance of collections in fields, orchards, plantations, herbal gardens, botanic gardens and more recently in genebanks. Vegetatively propagated crops, plantation crops and perennials are usually conserved in fields. Field gene banks have both advantages as well as limitations. In many cases, conservation methods involve *ex situ* storage as seeds backed by a limited number of clones in field or *in vitro* gene banks. National Bureau of Plant Genetic Resources operates through its national network as well as with its 30 co-operating centers. The primary aim is to regenerate, evaluate, characterize and document accessions stored in base collections of the National Gene Bank. Field maintenance and regeneration also provide opportunity for identification of donor parents which are likely to be used in breeding programmes. However, too frequent regeneration of germplasm can led to genetic erosion due to environmental hazards, genetic drift, selection and mechanical or human errors. Provision of medium term storage facility for a field gene bank can go a long way in reducing this risk.

Conservation and development of Gene Bank (Seed Repository) for long term conservation of germplasm of diverse agri-horticultural crop is considered to be the most appropriate method under *ex situ* conservation primarily because it avoids the need of too frequent regeneration in the field and all other attendant risks. The method is highly suitable to the majority of arable and horticultural crops. These are quite amenable to desiccation (seed drying upto 5% moisture content) and can be easily stored at very low temperature (–20°C) for considerably long time.

The Indian National Bureau of Plant Genetic Resources (NBPGR) formally established in 1976, but actually founded on activities initiated in 1946, is the nodal organization in India for exchange, quarantine, collection, conservation, evaluation and the systematic documentation of germplasm resources. NBPGR has already established India's 'National Gene Bank' with the national mandate for long term conservation of germplasm (base collections). This includes infrastructural facilities for long term storage (–20°C).

Tissue Culture repositories

The gene banks are capable of conserving genetic diversity of orthodox crop species for prolonged storage periods, while simultaneously ensuring their full genetic potential. Yet, there are several species known to possess recalcitrant seed behaviour which poses serious problems to their conservation. In addition, several species do not reproduce sexually and are propagated only through vegetative means. Some of them may be out-breeders and highly heterozygous.

Recently, tissue culture technology (*In vitro* culture technique) has grown rapidly as an aid to the conservation of germplasm. It is evident that single cell, suspension and protoplast cultures, callus, organs and meristems, shoot tip cultures, axillary buds, pollen, anthers and even embryos (both zygotic and somatic) can be preserved for varying lengths of time. The *in vitro* cultures can be preserved using minimal media (slow growth), growth retardants and through vitrification or osmotic regulations as well as by variation in temperature and light regime. The above mentioned culture systems can also be freeze preserved in liquid nitrogen (–196°C).

With recent advances in computer science, not only germplasm documentation systems are becoming more powerful and user-friendly, but also data exchange and the sharing of information among different systems are becoming easier. Sustained management, in short, is defined as "the management and utilization of forest areas in a manner which preserves their biological diversity, productivity, regenerative vigour, vitality and their capability to fulfill important ecological, economic and social function at the local, natural and international levels now and in future and does not harm the ecosystem (Tzschupke, 1998). While saving genes, species habitats are important, the need for a conservation framework is very important, which would enable saving biodiversity and using it sustainably and equitably. Agricultural, forestry and technological practices are now included, besides maintenance of protected areas. Conserving the genetic diversity of domesticated species is also as important as conserving wild species (Reid, 1992).

In situ conservation

Options for *in situ* conservation range from nature reserves from which all human intervention is excluded, through national parks in which economic activities with a potential to disturb the natural ecosystems are carefully regulated, to the implementation of special management regimes in areas used primarily for agriculture and forestry. The identification of specific areas in which a deliberate attempt is made to increase and maintain intraspecific diversity of key species, is another approach. The Man and

Biosphere (MAB) program of the United Nations Economic, Social and Cultural Organization is perhaps the largest coordinated global attempt to establish *in situ* reserves, one of the objectives of which is the conservation of natural areas and the genetic materials they contain. Under the MAB program, more than 250 biosphere reserves have been established around the world.

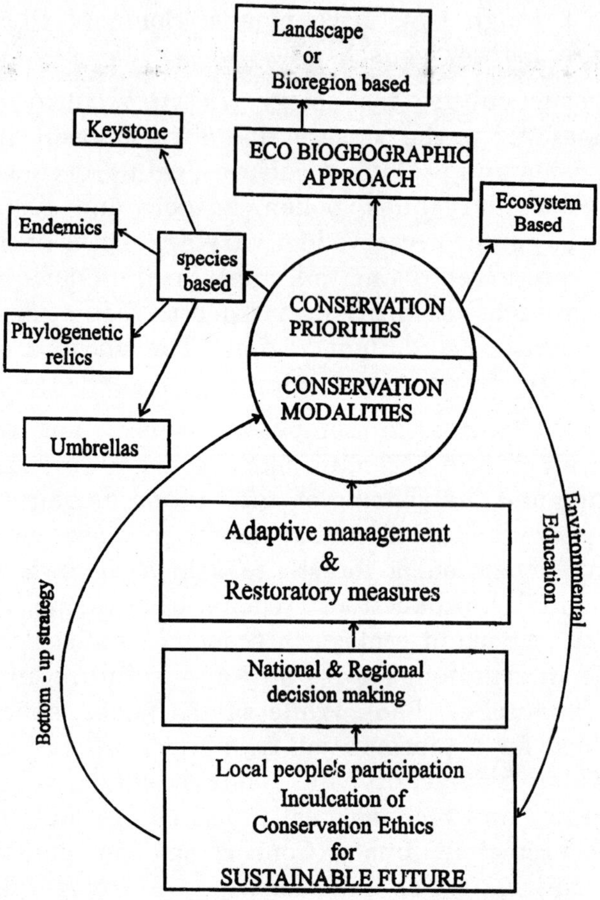

Figure 17.2. Conservation priorities and modalities.

As more attention is paid to *in situ* conservation, more innovative approaches are developed. For example, locally based conservation seeks to conserve biological entities at the farm, community, or regional level. Local issues such as traditional and cultural behaviour and knowledge play a large part in conservation efforts (Fig. 17.2).

There are 54 national parks and 372 sanctuaries with a total area of over 10 million hectares or 3.3% of the area of the country. These include 13 biosphere reserves for which detailed documents about their biological holdings were prepared before declaring them so. The biosphere reserves are:

Nilgiri (Tamil Nadu, Kerala and Karnataka), Namdapha (Arunachal Pradesh), Nanda Devi (Utter Pradesh), Valley of Flowers (Uttar Pradesh), Great Nicober, Gulf of Mannar (Tamil Nadu), Kaziranga (Assam), Manas (Assam), Sundarbans (West Bengal), Thar Desert (Rajasthan), Kanha (Madya Pradesh), Nokrek Tura (Meghalaya) and Rann of Kutch (Gujarat). One of these (Gulf of Mannar) is a marine biosphere reserve (Fig. 17.3).

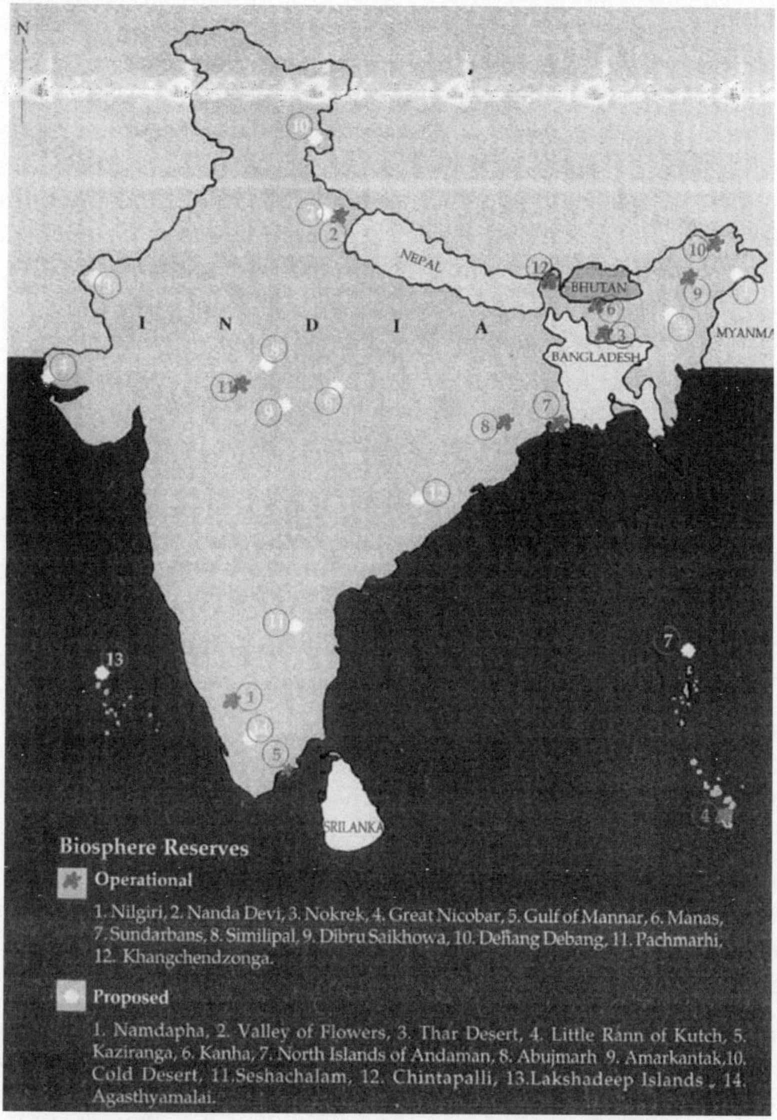

Figure 17.3. Conservation priorities and modalities.

Wildlife management is comparatively new to India and yet it needs no emphasis. Wildlife is an integral part of the world. The animal needs the forest and it is supported and protected by the forests during its life span. The forests also need animal life, though not to the degree that animals need the forest. In dealing with our problem to improve the status of wildlife, due care has also to be taken about the forests of our country. Though the nature, status and problems of our forests are complicated, efforts should be made to promote wildlife through the existing forests of the country, which should provide three distinct needs of its wildlife: (a) places to breed in safety, (b) places of refuge, and (c) adequate supply of food and water.

The dynamic phase of wildlife management is called 'environment manipulation", based upon biological principles. The general principles that govern the functioning of communities and the place of wildlife within them are the same all over the world, though the habitats, occupied by the wildlife are diverse. These animals establish themselves in habitats which provide them with the basic needs for their normal life, and different animals have different requirements for food, water, resting, place, escape cover, etc. If the animals are forced to live in a habitat that does not have all the requirements of their life, then natural selection will weed them out. However, any wildlife species according to its place in the diverse ecological niches, provides a useful basis for tackling problems regarding its conservation. A developing country like India, which still has abundant land and natural resources together with a favourable climate, continues to be a valuable heritage and potential for economic growth and progress.

There are three broad methods by which our existing wildlife may be preserved:

1. Most wildlife could be saved by sufficiently rigid protection of their natural habitat combined with controlled exploitation of species which are of economic importance.
2. Some wildlife may be driven away from their natural habitat so that their requirements are in conflict with human interest. The only possible course for maintaining their valuable number is national parks, sanctuaries, game reserve, where they could thrive unharmed.
3. In the case of the most endangered species which have reached a point of extinction in their favourable habitat, their survival may be ensured through the maintenance of breeding stock in zoological parks in suitable parts of the country.

In attempting to introduce conservation measures, the Fauna Preservation Society and the Bombay Natural History Society was established in 1885. This was followed by the Wild Birds and Animals Preservation Act in 1912. Subsequently came the Indian Forest Act of 1972, the Bengal Rhinoceros Preservation Act of 1932 and the Rhinoceros Preservation Act of 1954.

The most important modern development was the foundation in 1948 of the International Union for the Protection of nature which was subsequently renamed as the International Union for the Conservation of Natural Resources (IUCN). This organization coordinates the work of preservation societies all over the world, besides carefully watching the performance of species threatened with extinction in order to offer suitable advice to the concerned Government. In August 1949, the Survival Service Commission was inaugurated with the undermentioned programme to (a) collect all available information about the threatened species, (b) verify this information through questionnaire from the concerned area, (c) give wide publicity regarding such species, and (d) advise and assist the concerned Governments including other organizations of the area. In September 1958, the responsibilities of the IUCN were further extended establishment of an International Committee on National Parks, with a view to strengthen International co-operation regarding reserves and national parks. Its purpose is to serve as an information centre concerning reserves and parks, and to undertake scientific research.

In December 1952, the Indian Board for Wildlife was inaugurated. The function of the IBWL is: to (a) advise the states for the conservation and control of wildlife through co-ordinated legislative and practical measures, (b) sponsor the setting up of national parks, sanctuaries and zoological gardens, (c) promote public interest in wild life and the need for its preservation, (d) advise the Government on policy in respect of export of living animals, trophies, skins, fur, feathers and other wildlife products, (e) prevent cruelty to birds and beasts caught alive, and (f) perform such other functions as are germane to the purposes for which the Board was constituted. Later in 1959 the wildlife Preservation Society was formed at Dehradun.

The IBWL has defined a national park as 'an area dedicated by statute for all times to conserve scenery, natural or historical objects of national significance and wildlife and where provision is made for the enjoyment of the same by the public". A sanctuary on the other hand is "an area where killing or capturing of any species of animal is prohibited except under orders of the competent authority and whose boundaries and character should be sacrosanct as far as possible".

Conservation DO^{S} and DONT^{S}

Like aquatic and land resources a park/sanctuary may be large or small, but the planning should be based on a scientific background. To provide the basic need for the wildlife the following steps may be taken as far as practicable for the conservational purpose.

(a) The forest inside the national park/sanctuary should be left unexploited as far as possible. It would be an ideal environment for the wildlife to get the fullest measure of protection, and in such preserved conditions of life they multiply freely.

(b) Poaching needs to be ruthlessly prohibited in the sanctuaries in particular and generally in other areas outside them, where wildlife still exists. Mobile armed units should be attached to all sanctuaries to patrol regularly the entire sanctuary to stop illegal activities.

(c) The water holes and stream sides should be managed and guarded carefully for the benefit of wildlife. Such critical areas may be of small size, but their importance is disproportianetly great because they supply important elements in wildlife ecology for larger areas during the host months of the year.

(d) Provision of dust bath for animals should also made by leaving some exposed soil for their dusting. These spots may be prepared in the form of small mounds elevated a few inches above the surrounding forest floor to allow for drainage and to dry quickly in wet months.

(e) Some artificial salt-licks are also to be provided in the sanctuaries since the wildlife need them at regular intervals for their normal health. Such salt-licks are necessary for the animals especially in area where natural salted clay is scanty.

(f) Concentrated grazing by domestic livestock is inimical to the interest of wildlife, since, the wildlife can never compete successfully with domestic stock under ordinary circumstances, and (b) domestic animals may be responsible for transmitting contagious diseases like foot and mouth disease, rinderpest, surra (sleeping sickness), haemorrhagic septicaemia, anthrax, etc. During 1958-1961 the Indian wild ass in the Rann of Kutch was suspected to have suffered a heavy loss in population by surra and with this disease again two wild elephants were reported to have died in the Quilon district of Kerala in 1973. In 1968, rinderpest occurred among the gaur in Bandipur sanctuary and within a few months not even a single animal could be sighted. Sambar and cheetal too were seriously affected and over hundered wild elephants were believed to have died in the North Cachar Hills during 1949 and several rhinos reported dead in Kaziranga around 1944 due to anthrax. In Sariska over thousand sambars died within a year in 1970-71, by haemorrhageic septicaemia.

In India there are over 150 million cattle and domestic buffaloes and 100 million gats and sheep of which at least 10 per cent graze exclusively in the forests where our wildlife survives. Under the above circumstances all domestic livestock grazing should be prohibited inside the sanctuary.

(g) Cultivation near the sanctuary should be avoided as far as possible since it may have an indirect effect on the wildlife. The chemical control, for example, of certain pests is often desirable in the interest of agriculture, but these pesticides can present ecological hazards for wildlife.

(h) Scientific studies of the wildlife sanctuaries, particularly of threatened species, by qualified persons should be encouraged in order to improve the status of the animals even by breeding them in captivity to rehabilitate them in suitable habitats.

(i) Educational and publicity programmes should be undertaken to create public opinion in favour of wildlife. There is a general lack of knowledge about the conservation of nature and value of wildlife in India.

(g) The third category pertains to the most endangered species which have reached a stage of extinction. The only hope for survival of the most endangered species is through a successful captive-breeding programme. In India, the zoological gardens may play an important role in this programme. So far the Delhi Zoo has successfully bred, among others, the Nilgiri langur, lion-tailed macque, brown antlered deer, Indian otter, great Indian one horned rhinoceros, tiger lepard and sloth bear. Similarly Nandankanan Biological Park has recorded the birth in captivity of the spotted deer, sambar, barking deer, hog deer, nilgai, tiger and leopard. The Mysore Zoo has produced captive breeds of Indian elephant, Indian lion, and tiger. The Alipore Zoological Garden, Kolkatta has successfully bred white tigers and Indian lions and has had the privilege of the first captive breeding stock of the most endangered species of Brown antlered deer. Now an effort is yet to be made to rehabilitate such captive breeds in their natural habitat. This has to be carried out in a gradual manner.

Today the most encouraging fact concerning the future of animal life of India and the world is that everywhere there are bodies concerned with preservation and conservation so that they are steadily gaining public support for their work. Public opinion has mostly been in favour of protection of rich natural forests from being sacrificed at the altar of power projects.

Introduction of exotic species should be carefully done after a thorough study of all aspects of ecology of the proposed exotic species, so that the earlier experience of the Indian mongoose introduction into Puerto Rico or the rabbits into Australia may not be repeated.

Another important aspect of management pertains to predator control, which as is at present understood by ecologists, considers the predator to be an essential part of the ecosystem. The number of predators at the apex of a food pyramid is dependent on the number of prey animals at lower levels, never the reverse. Predators may severe a useful role in keeping the prey population within limits imposed by the carrying capacity of its habitats. Lack of predator pressure might release a population explosion resulting in habitat detoriation and culminating in massive mortality caused by starvation and disease. Suitable cover, whether it be a swamp core, tall grass cover, or a leafy canopy is essential to protect wildlife from predators.

Participatory Forest Management (PFM)

PFM has earned recognition as an important measure of conservation strategy, wherein the state and the community become partners in the management of the forest resource. The state continues to own the resource, but the benefits are shared. The community develops an economic state in the preservation of forests leading to conservation and sustainable use. The experience gained in India during the last 25 years shows that the process of natural forest degradation can be reversed through PFM and that forests can provide the local community no-wood forest products on a continuous or seasonal basis. Since management includes conservation, sustainable use and equitable sharing of benefits, the ultimate outcome will result in the conservation and enhancement of natural resources. Since the non-involvement of local communities in the past has resulted in a severe depletion of forest resources in India. PFM will play an important role in protecting prime forests or regenerate degraded forests (Swaminathan, 2000). To quote swaminathan, "By recognizing that conservation efforts represent a continuum, with rural and tribal families performing a vital function in preserving precious genetic diversity in important plants and farm animals, we will be able to attend to all links in the conservation chain. Community conservation is a value-added link in the conservation system, since local families not only conserve, and add value to the conserved material through solution and information".

Increasing realization that genes, species, ecosystems and traditional knowledge and wisdom that are being lost at an increasingly accelerated pace, the involvement of PFM adds another dimension to conservation efforts.

CHAPTER - 18

INTELLECTUAL PROPERTY RIGHTS

Intellectual property deals with creations of the human intellect. It encourages innovation and inventive activity. Confidence in intellectual property protection is a powerful stimulus to innovation; such protection provides the necessary incentive for scientific, technological and intellectual creativity. The protection of intellectual property rights enables countries to participate more actively in international trade and influences investment decisions. Developing countries have, more particularly since the last decade or so, come to recognize that economic development is becoming increasingly technology-based and is accelerated through the use of new high sophisticated technologies. Intellectual property signifies this advancing knowledge in the form of new ideas, techniques, processes and products having economic and commercial potential. Over the past decade and a half, industrial property has acquired an increased prominence at both the national and international levels. We see in developing countries, the policy-makers, industrialists, entrepreneurs and academics paying increasing attention to the importance of effective protection of intellectual property assets.

Intellectual property consists, in the main, of two aspects, namely: (i) *Industrial property*, largely in respect of technological inventions, trademarks for goods and, services, industrial designs and (ii) *Copyright*, largely in literary, musical, artistic photographic and audiovisual works. In recent times the term intellectual property is used to include all property resulting from the exercise of the human intellects, with such aspects as utility models, plant variety, geographical indications, etc. (Alikhan, 1994).

The principal objective in protecting intellectual property is, again in the context of the development of developing countries, to encourage creative activity and to provide to the public easily, economically and speedily, the result of such activity.

Encouragement of creative activity requires not only the recognition of the creators *per se*, but providing for them a means of obtaining a reward for their ideas and inventions. The creator, be it an individual or an enterprise, has also to be accorded the right to prevent others from using his, or her, or its, ideas without consent and without compensation or remuneration being paid to the individual creator or the concerned enterprise. The availability of such an exclusive right also provides legal security as well as the necessary fillip to scientific and technical institutions and enterprises that are wishing to do so, to encourage, through material resources and the necessary funding, to use skills in research and development of worthwhile new ideas that can be utilized in improving existing products or in making new products.

Intellectual property protection has helped to guide and evolve programmes and activities in business and university research, and has facilitated technological development. Well enforced intellectual property protection has often led to shifts in the costs linked with technology development from the public to the private sector of the economy. Again, where confidence in intellectual property protection prevails, links have strengthened between business and the universities; there is a greater need for industry and business to turn to university researchers for various consulting requirements. On the other side, university researchers, through their increased contact with private business and industry research, get the required stimulus for their work. The intellectual property system acts as a "facilitator" for the framework of a stimulating relationship between universities and business.

Developing countries would need to promote such interaction. They would also need periodic re-examination of their fiscal, industrial, trade and industrial property policies and the interaction between them in the context of seeking and retaining export markets and directing their production patterns to this end. Weak internal linkages that inhibit the efficiency of the productive sector will require to be corrected.

IPR AND TRADITIONAL KNOWLEDGE

Traditional knowledge protection and promotion

The process of innovation also includes the often unrecognized technology innovation that takes place in an informal system of innovation, be it by artisans, farmers, tribes or other grassroot innovators. Indeed many societies in the Third World have nurtured and refined systems of knowledge of their own, relating to such diverse domains as geology, ecology, botany, agriculture, physiology and health. These informal innovators have, therefore, generated such a rich store of traditional knowledge (Mashelkar, 2001).

IPR and traditional medicine

Traditional medicine (TM) plays a crucial role in healthcare and serves the health needs of a vast majority of people in developing countries. Access to 'modern' health care services and medicine may be limited in developing countries. TM becomes the only affordable treatment available to poor people and in remote countries.

World health organization (WHO) defines traditional medicine as the sum total of all the knowledge and practices, whether explicable or not, used in diagnosis, prevention and elimination of physical, mental or social imbalance and relying exclusively on practical experience and observations handed down from generation to generation, whether verbally or in writing. Health care providers worldwide including major pharmaceutical giants are turning to incorporate many of these into their mainstream activities. As traditional medicines are largely based on medicinal plants, indigenous to these countries, where the system has been in vogue for several centuries. The effort is on accessing them either directly or through the use of modern tools of breeding and cultivation, including tissue culture, cell culture and transgenic technology. IP issues linked to such endeavors remain unresolved.

The protection of TM under intellectual property rights (IPRs) raises two types of issues. First, to what extent it is feasible to protect, existing IPR system. Certain aspects of TM may be covered by proposals to develop *sui generis* systems of protection. Such proposals are based on the logic that if innovators in the formal system of innovation receive a compensation through IPRs, holders of traditional knowledge should be similarly treated.

As a corollary, an intellectual property "culture" needs to be assiduously and deliberately promoted, which would encourage innovative and inventive activity linked to market needs, encourage scientific and technological creativity; help modernize intellectual property infrastructures.

Industrial Property: The Patents System

As is known, an invention, which the patents system seeks to protect, is a novel idea that permits, in practice, a technological solution to a specific problem in industry. The idea, to be patentable, should be new, non-obvious, and applicable in- industry, in the sense that it can be used to make a product. An inventor, be it an individual or an enterprise, has, in filing an application for registration of an invention, to disclose and describe it. The concerned government authority-usually the Patent Office, after the necessary search and examination, issues a document or a patent that describes the invention and declares the inventor or enterprise as the owner. The issuance of a patent creates a legal situation in which the patented invention can normally be exploited, that is made, used, sold, imported, only with the authorization of the patentee. While this protection of an invention is limited in time (generally 15 or 20 years), the disclosure itself provides an important source of technological information. Considering that over one million patent applications are filed each year throughout the world, the patent system is a veritable storehouse of valuable information (Subbaram and Thyagarajan, 1998).

Effective intellectual property protection plays an important role in attracting technology transfer and in obtaining the desired range of technical knowledge available in other parts of the world. The introduction of the latest and newest technologies in our productive sector plays a critical role in economic growth, as well as in the effort to increase per capita national income.

Adequate legislation and enforcement of intellectual property rights undoubtedly helps in the transfer of technology as it assures foreign investors and foreign licensors that their technology will not be exposed or revealed to competitors. In other words, a barometer for the easier availability and supply of technology and know-how to industry in developing countries is the existence of adequate and effective patent protection.

Copyright

Copyright provides for protection of literary and artistic works. This includes *literary works* such as books, short stories, poems, other writings, instruction manuals, catalogues; *dramatic works,* including plays, films, scripts, scenarios and other intended to be performed such as *choreographic works, musical works* including both light or serious scores, combination of melodies, operas, orchestras; *artistic* works including paintings, sculptures, engravings, maps, drawings (sketches, architectural drawings), *photographic* works (portraits, landscapes, etc.) and *audiovisual works* including documentaries, television programme, newsreels, videotapes, video cassettes, etc.

An important function of the patent system, as was referred to earlier, is to provide access to valuable technical information. This helps to develop national technological innovation on the already known state of the art and assists, through examination of the already existing technical knowledge, in decisions to launch new R & D projects.

Industrial Property: Patent Information

The patent system serves, as has been mentioned, for invention and innovation through the resource of technological information contained in disclosed patent documentation. Effective use of this documentation is an aid in the transfer of technology. The technology disclosed serves to stimulate ideas for further invention and innovation. The economic value of patent information is that it provides industry with technological information which can be used for commercial purposes.

As will be seen from what has been discussed, protection of intellectual property rights is increasingly becoming a factor to be reckoned with in the process of technology transfer, technology growth, industrial and socioeconomic development, international trade, identifying of new markets in export trade and their retention, promotion of national inventive and innovative activity, assisting the expansion and growth of small and medium-sized industries, and in a determined bid to increase links between them and university-based research.

In general, patent law requires that for an invention to be patentable, the following criteria are to be followed:

A. The invention must be new and should not be known anywhere in the world on the date of filing of the application.

B. The invention must involve an inventive step.

C. The invention must be industrially applicable i.e. should be useful to society.

The General Agreement on Trade and Tariffs (GATT) formed in 1947 is a forum for stabilizing international trade rules. The trade related aspects of intellectual property rights (TRIPS) under the GATT provisions are: duration of patents to be uniformly 20 years from date of filing; availability of patent rights are locally produced or imported; providing protection for microorganisms and microbial process and enhancement of an effective system for the protection of new plant varieties. There are differences in polices and regulation and in India the patent law permits only 'process' protection in food, pharmaceutical and chemical sector and not product protection. The term of the patents is only 7 years. The provisions of TRIPS with agreement exclude "plants and animals" from patentability, but makes it obligatory to provide patent protection for 'microorganisms' and microbial processes.

A third amendment has been made to India's Patent Act, this time extending the country's patent protection to include agro-chemicals, food, and biotechnology products, apart from drugs and pharmaceuticals. This was made in order to meet with the World Trade Organization (WTO) Trade-

Related Aspects of Intellectual Property Rights (TRIPS) Agreement, criteria for patent rights, namely novelty, inventive step, and utility; however, it has not defined microorganisms and microbial processes. Thus, the new patent regime will cover seeds created by novel means, including transgenic seeds; and claims in gene patent applications may pertain to genes or partial DNA sequences, proteins encoded by these genes, vectors used for transfer of genes, genetically modified micro-organisms, cell, plants, and animals, and the process of developing a transgenic product. The amendment went into effect on January 1, 2005.

Case studies on patents include :—

A. Methods for producing transformed cotton cells by tissue culture and regeneration of intact, sexually mature cotton from the genetically transformed cells. This invention has novelty, an inventive step and utility.

B. Use of turmeric in wound healing. The turmeric case is a landmark wherein for the first time a patent based on traditional knowledge was resolved.

C. Method of controlling fungi on plants by the aid of a hydrophobic, extracted from neem oil.

D. Case of re-examination of US patent on Basmathi rice.

The Biological Diversity Bill

The biological diversity bill has been based broadly on the objective of CBD with special focus on the aspects of equitable use of benefits arising out of sustainable use of biodiversity and traditional knowledge. The salient features of the Bill are to: regulate access to biological resources of the country with the purpose of securing equitable share in benefits arising out of the use of biological resources and associated knowledge relating to biological resource (ii) conserve and sustainable use biological diversity (iii) respect and protect knowledge of local and indigenous communities related to biodiversity (iv) secure sharing of benefits with local people as conservers of biological resources and holders of knowledge and information relating to the use of biological resources(v) conserve and develop areas important from the standpoint of biological diversity heritage sites (vi) protect and rehabilitate threatened species and (vii) involve institutions of self-government in the broad scheme of the implementation of Act through constitution of communities (GOI, 2000b).

The bill in general is more of a regulatory one, but contains the following important clauses pertaining to protection of IPRs of traditional communities and benefits sharing:

A. People's knowledge shall be protected through registration at local, state and national level by a *sui generic* system of IPRs (Clause 36 (4) & 41).

B. Any person applying for IPR in India and abroad relating to biological resources occurring in and accessed from India shall obtain prior permission of national authority and abide by the benefit sharing conditions imposed by the authority (Clause 6).

C. The national authority shall oppose the grant of IPRs worldwide relating to biological resources or knowledge derived from India (Clause 18 (4)).

D. No foreign agency can access biological resources occurring in India and related knowledge without prior informed consent of national authority (Clause 3).

E. The national Biodiversity authority in consultation with local bodies shall impose terms and conditions for securing equitable sharing of benefits. Monetary benefits shall be deposited to the National Biodiversity Fund, except in cases where the biological resources and knowledge are accessed from specific individual or group of individuals, in which case the monetary benefits shall be directly made to the providers (Clause 21). The national fund would be used to reward people for their conservation efforts and knowledge as claimed by the village level management council (Clause 41).

Bioprospecting

Biotechnology (BT), Information Technology (IT) and Herbal Technology (HT) are emerging areas of research and development in the current century. Biodiversity and traditional knowledge systems offer the key resources for these promising enterprises. The prospects of biodiversity for new medicines, foods, insecticides, pesticides and other commercially valuable genetic and biochemical products and processes are booming, thanks to recent advancements in biotechnology and genetic engineering. The exploration of biodiversity for commercially valuable genetic and biochemical resources is termed as *bioprospecting* which is an important component of major bioprospecting programmes. Transgenic technologies now facilitate transfer of genes of desirable traits or chemical constituents from animals or plant to bacteria. The resultant transgenic bacteria are transformed into chemical factories for providing desired products such as enzymes, drugs and pharmaceuticals. Indefinitely, isolating and characterizing novel bioactive compounds have opened up new opportunities in natural products research. Bioprospecting can therefore bring forth substantial economic returns for the products and processes to be derived from biodiversity and biotechnology. Among bioprospecting programme should besides economic benefits, contribute to fund conservation, inventorying and monitoring biodiversity (Pushpangadan, 2002). Any bioprospecting programme should, therefore, be carried out with the end in view that, apart from direct economic benefits, such activities that would contribute directly or indirectly to fund conservation, inventorying and monitoring. The need to protect IPR and its harmony with world trade cannot be overemphasized. Given the global trends in capturing the intellectual property markets, the Third World nations like India now need to look ahead for the best possible ways and means by which they can generate IPR and build up IPR covered bioindustrial regimes.

In order to understand the full implications of intellectual property rights on agricultural production, one should also understand how 'plant breeder's rights' and patents function. A distinction between plant breeder's right and patents should also be fully appreciated. As discussed earlier, plant varieties are excluded from the list of patentable articles. Plant varieties are generally protected in several countries (not in India) through *plant breeder's rights* (PBR) or *plant variety rights* (PVR). Through these rights, further propagation of the variety is restricted. PBRs provide a more flexible model for developing countries allowing more flexibility in respect of the development approach. PBRs were created to satisfy the needs of large plant breeders who wanted to recoup their investment in breeding successful, commercial plants. The farmers' privilege supports farmer's livelihood and allows them to innovate further.

Conclusion

Following establishment of the international institutional mechanisms, such as, the Convention on Biological Diversity (CBD) and the WTO, and further, the signing of International Treaty on Plant Genetic Resources for Food and Agriculture (ITPGRFA), the growing importance and the global scope of IPR in agriculture are well realized and recognized. The IPR, after long debate, is recognized as an asset and means of rewarding and harvesting the fruit of agricultural research and development. Reorganization of intellectual property right provides an effective means of protecting and rewarding innovators. This acts as a catalyst in technological and economic development. The essence of regulation of IPR by law is to balance private and public interests.

CHAPTER - 19

BIODIVERSITY: THE FUTURE SCENARIO

Man derives many important services from communities, ecosystems and landscapes, which are the vehicles for individuals, populations and patterns of process. Insects provide many natural services such as natural control of pests, genetic resources for the biological control, pollination of crops, honey production and the like. Concern for ecosystems is therefore of great importance for all its individuals, species and populations. Concern for individuals, populations, species, assemblages, communities and ecosystem comes about by protection, appropriate management and awareness. One of the beneficiaries of biological conservation in taxonomic research, a dying science for decades, is regaining the attention of governments and funding agencies.

With a better understanding of the economic development potential of natural assets, the wisdom of setting aside protected areas are laudable. Systematics is linked not only to diverse aspects of biology, but also to the fortunes of the international conservation management which is concentrating its attention on the threatened environments of the Tropics (Wilson, 1984). Wilson further states that while ants and termites make up 1/3 of the biomass in tropical forests, there are only 8 entomologists worldwide who are competent to handle them !

People today are increasingly conscious of the direct cost of loss of genes, genera and species and ecosystem conserving nature's diversity. Besides, developed and developing countries, scientists, non-governmental organizations and the private sector have to appreciate the relevance of biodiversity to their own needs. It is certainly a positive sign that local communities are defending conservation of biodiversity and one of the best examples is the 'chipko' movement where village women in Northern India literally hugged trees to protect them from destruction. The three driving forces behind this conservation movement has been science, Biotechnology and grass root activities. In other words the fragmentation of biodiversity, cryopreservation, restoration ecology, captive breeding and deigning resources are among the most important aspects of conservation.

Biotechnology has opened up newer avenues for greater use of biodiversity and genetic diversity is the key raw material in agricultural research bringing the entire world's biota within reach. Commercial demands for biodiversity are on the increase by actively seeking to identify new economic values of biodiversity. But one has to be overcautious, since breakthroughs in genetic engineering could ultimately curtail the demand for natural source of diversity. It cannot be denied that the loss of species will entail ecological risks, as for importance, the loss of keystone species can dramatically change the structure and function of biological communities.

Multidimentional linkages in Biodiversity

Ecology and environmental biology have blossomed over the last century, leading to new insights with the generation and maintenance of biodiversity at all levels of organization. With increasing trends of the relationship between organisms and their environment, the interaction between organisms and evolution of new life forms, besides factors that lead to their extinction or preservation are of paramount importance. High quality in target courses on biodiversity is important, since biodiversity is no longer the concern of scientists alone, but also of managers in industry, students and teachers. Citizens can form what could be called a *conservation corps*. Regional biodiversity strategy should be an aspect of management action plan, restoring an environment such as a stream, declaring an area as wild etc. Baseline monitoring to measure a set of chemical and biological variables so as to be safe, is equally important.

One of the conspicuous challenges to biological conservation strategies in the present decade is a shift in focus from undisturbed natural habitats, including protected areas that cover only a very small proportion of the world land area, to the 'managed ecosystems' which include agricultural systems, forestry systems and human settlements, contributing to 95% of the terrestrial environments. Crucial to a check of the unprecedented biodiversity erosion in such 'managed ecosystems', is a proper understanding of intricate, multidimentional linkages among the basic components of biodiversity, viz., the productive biota. For instance, where natural enemies or more effective biological control agents inhabiting wild vegetation reservoirs at field edges, and in association with crops, an agroecological and landscape approach is rather indispensable to promote sustainability in agroecosystems. Agroecological practices are oriented towards development of agroecosystems with a minimal dependence on high agrochemical and energy inputs. By assembling a functional diversity, it should be possible to initiate synergisms which support agroecosystem processes by providing ecological services such as the activation of soil biology, the recycling of nutrients and the enhancement of beneficial arthropods. In other words, pollinators, natural enemies, earthworms and soil microorganisms are all key biodiversity components that play important ecological roles, thus mediating such processes as genetic introgression, natural control, nutrient cycling, decomposition etc. The key factor in this context is the identification of the type of biodiversity that is desirable to maintain and/or enhance in order to carry out ecological services. The determination of the best practices that will encourage the desired biodiversity components, in the context of changing spatial and temporal diversity profiles, associated with changing cropping patterns and agroclimatological conditions, also becomes very relevant.

The rice field ecosystem which provides one of the most sustainable examples, with an amazing spectrum of floral, faunal and microbial diversity can be considered as "opportunities to the drastic condition in these temporary wetlands". The diverse soil benthos of rice field ecosystems (oligochaetes) maintain soil fertility by nutrient recycling and several arthropod natural enemies (predators and parasitoids) of rice pest insects and predatory aquatic insects of vector mosquito species, form an array of natural biological control agents. In its aquatic phase, a rice field is a rich and productive biological system which can produce a crop of aquatic organisms, both plant and animal for human consumption, besides the crop of rice. An understanding of the ecology of aquatic organisms and the biological cycles of the aquatic phase of the rice fields is a necessity if it is to be exploited for the concomitant production of rice and fish.

Hence, such agrolandscapes should be properly managed to increase species diversity within landscape patches to conserve the genetic material and to maintain local cultivars and land races. Diversity studies should include analysis of taxonomic, genetic, ecological, phylogenetic and biogeographic aspects of species complexes in the agroecosystems. Today, biodiversity is considered as a fundamental principle in agricultural sustainability and a stockpile for biotechnology for future studies. It should focus on biodiversity as an organizing principle especially in the management of agroecosystems. A shift in emphasis from Integrated Pest Management (IPM) towards Integrated Biodiversity Management (IBM) appears quite warranted, if we hope to reap the benefits of agroecological approach in agroecosystem design and management.

The Tsunami Disaster: Biodiversity Implications

Earthquakes on the sea floor are the origin of the tsunami waves, inflicting colossal damage to human, animal and plant lives. The dynamic waves which are shock embedded, travel at speeds of over 1000 km per hour. The ocean bed rising more than ten miles and displacing hundreds of cubic kilometers of the overlying water generally a massive tsunami, striking coastlines across the India Ocean with walls of water as high as 10 meters (Fig. 19.1).

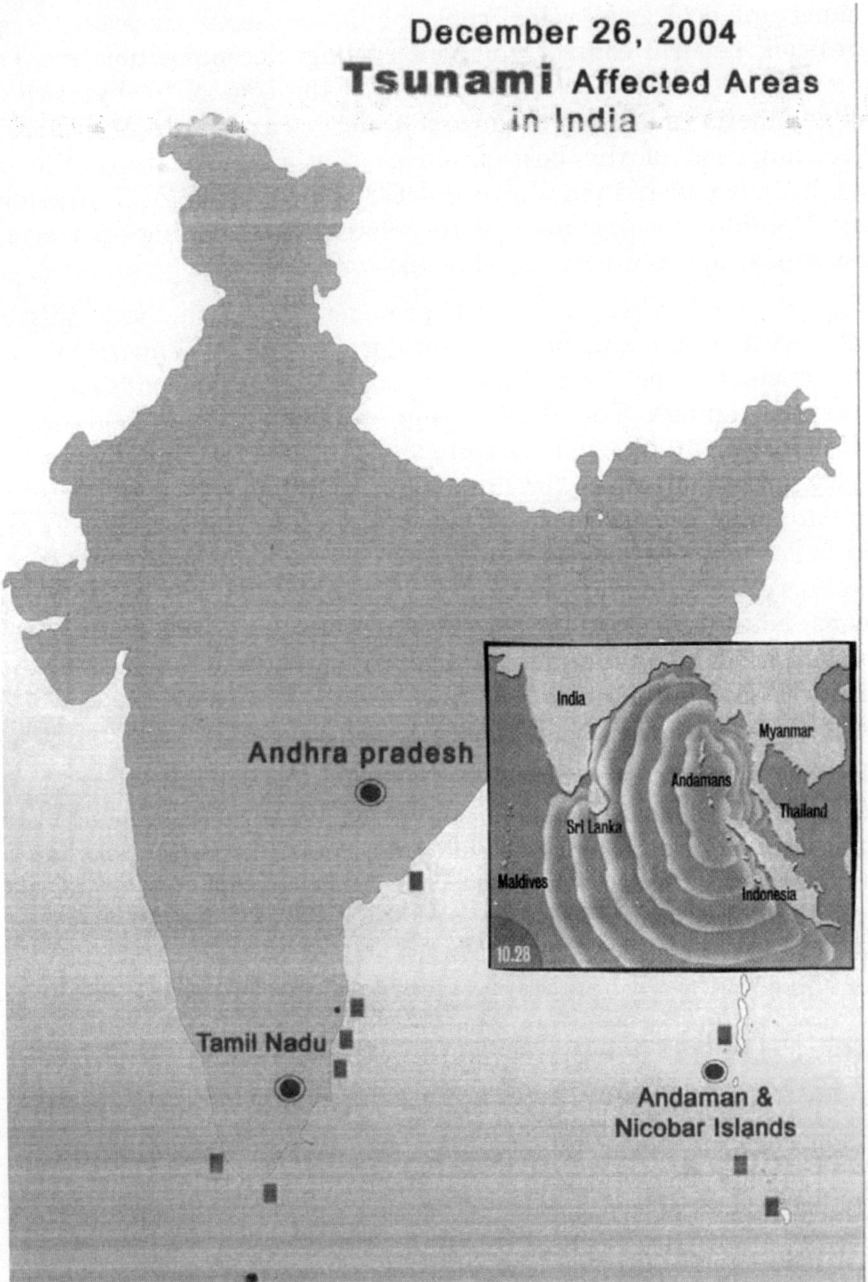

Figure 19.1. The Tsunami affected areas in India

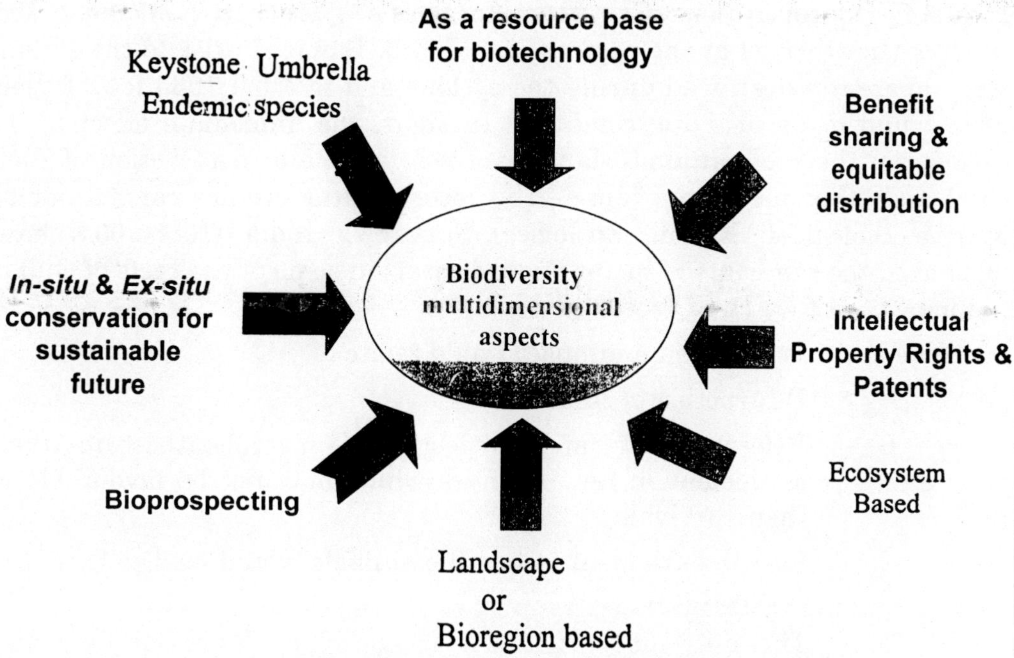

Figure 19.2. Multifunctional aspects of Biodiversity

The powerful tsunami waves struck the south east coast of India especially Tamil Nadu and Pondicherry and many islands of Andaman and Nicobar on 26.12.2004. Tamil Nadu coast, southern most part of Kerala and the entire Andaman and Nicobar were most affected. The ferocity of the waves has caused devastating and irrepairable damage to the marine biodiversity. Since damage assessment has yet to be done, it is easy to visualize the effects, through the impact caused by 1964 cyclone which submerged Dhanuskodi.

In general, the coral reefs were found to be affected by the tidal waves caused by the cyclonic winds than those of the Gulf of Mannar. The branching corals, especially *Acropora* spp. were the worst affected by the tidal rampage. Several colonies of *A. corymbosa* were found to be broken at their base and washed to the shoreward side of the reef and the lagoon in Palk Bay. The larger colonies of *A. surculosa* sustained similar damage at Manauli Island and several of them were noticed to be turned upside down. The massive forms like *Favia, Favites, Platyguara, Symphyllia,* etc. were however not much affected in the Gulf of Manar side. But there was an overall reduction of the living colonies of these massive forms in Palk Bay mainly due to the killing of them by the large quantity of sand deposited by the waves. The large platforms of *Montipora foliosa* and *Echinopora lamellosa* at Krusadai Island and Manauli Island were apparently little damaged. The alcyonarians also suffered severe loss. The dense formation of *Lobophytum* and *Sarcophytum* that thrived at the eastern end of Palk Bay reef prior to the cyclone was apparently destroyed during the cyclone and in June 1965 few of them were found to be in living condition. In short, the immediate effect of the cyclone on the reefs around Mandapam was a general destruction of their coral and other associated fauna. The recent status studies carried out by Marine Biological Station, Zoological Survey of India (1988-2004) have visualized the remnants of heaps of dead corals in some of the reefs of Gulf of Mannar.

The anticipated Tsunami impact would relate to:-

A. Destruction of coral reefs.

B. Relocation of many species of macrobenthos to their settlement in new habitats which may not be favourable to their survival.

C. Larval settlement of benthic animals would lead to their low survival rate.

D. The nutrient enrichment triggered by Tsunami may also induce a "crown of thorns starfish phenomenon"

E. The effect of Tsunami waves would bury most of the epibenthic molluscs affecting their contribution to reef building.

F. Population density of many finfishes would be affected by the waves which would bring sediments to the surface, which would asphyxiate them leading to mass mortality. Many other species of fish are bound to be affected by changes of water quality. The nutrient levels in water may increase to an alarming level and trigger red tide blooming which may also cause mass mortality of finfishes.

G. The breeding success of turtles may be affected.

The population of endangered Dugong reported from East coast especially from Gulf of Mannar and Andaman and Nicobar Islands may dwindle due to the reduction in the availability of food (=seagrass). There may be a change in breeding ground. The reduction in fish population due to tsunami may reduce the population of dolphins.

Coral reefs offer front line defence to the coastline. The tsunami waves would have caused the most serious damage to the coral reefs of Gulf of Mannar and Andaman and Nicobar Islands. The most delicate branching corals like *Acropora*, *Montipora*, etc. would have been destroyed by the waves (Fig. 19.3). This destruction would have also killed many species of invertebrates like crabs, stomatopods and pistol shrimps etc. which are associated with these corals. These corals not only harbour invertebrates but also serve as a refuge for a myriad of fauna and flora including reef fishes. Now the destruction would have left many reef animals homeless. Thus this may have a great impact on their population, which may decline due to predation. The demise of branching corals may pave way for the succession of hardy massive corals like *Porites*, *Favites*, etc. This may set in an ecological succession which may take many years for a new species composition to emerge. The delicate branching corals may need more than ten or above to re-colonize.

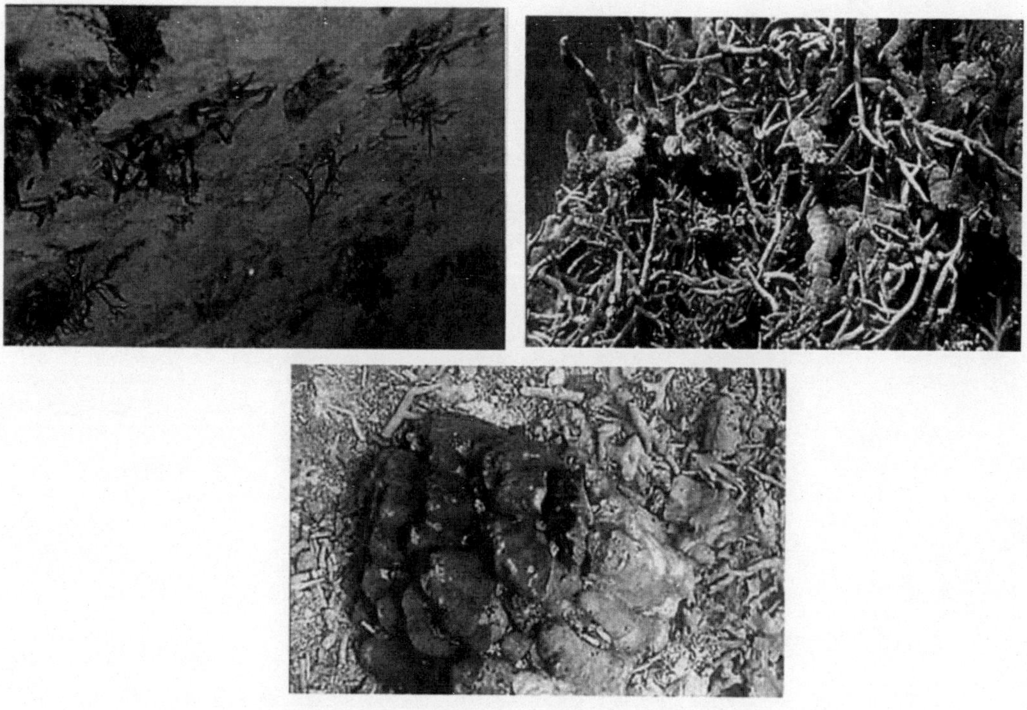

Figure 19.3. Coral reefs affected by Tsunami

Need for Management

The need for a natural disaster Management Authority to minimize the damage level cannot be overemphasized. This would be a long term programme to predict earthquake both on land and in the seabed sufficiently in advance to provide adequate waving and minimize loss of life. Such unforeseen calamities like Tsunami have once again confirmed the need for anticipatory research programmes to cope up with problems through adequate disaster management programmes such as introduction of mangroves preventing flooding and protecting biodiversity along the shores. To quote Prof. Swaminathan, "the strengthening of the ecological foundation for sustainable livelihood should be given priority. The rehabilitation of degraded mangrove forests and erection of 'bioshields', along the coast line should get due attention. The unique bioshields will consist of planting appropriate mangrove species as well as other salt tolerant plants. The 'bioshields' programme must include raising community nurseries of these salt-tolerant plants to provide livelihood security to the coastal population".

References*

Abrahamson, W.G. 1989. *Plant-Animal Interactions.* McGraw Hill Book Co., New York.

Alikhan, S. 1994. *Intellectual Property, The Developing Countries and Economic Development.* Rajiv Gandhi Institute, New Delhi, 82 pp.

Altieri, M.A. 1994. Biodiversity and Pest Management in Agroecosystems. Howarth Press, New York.

Altieri, M.A. and Nichols. 2002. Invasive arthropods and insect outbreaks in the context of ecology of mechanized agricultural system. *In.* G. J. Hallman and Schwalbe, C. P. (eds.) *Invasive Arthropods in Agriculture.* Oxford & IBH, New Delhi.

Altieri, M.A., Van Schoonhoven, A. and Doll, J.D. 1977. The ecological role of weeds in insect pest management system: a review illustrated with bean (*Phaseolus vulgaris*) cropping system. *PANS*, **23,** 195-206.

Ananthakrishnan, T.N. and Viswanathan, T.R. 1976. *General Animal Ecology.* Macmillan, New Delhi, 324 pp.

Ananthakrishnan, T.N. 1986. *Dynamics of Insect-Plant Interaction.* Entomology Research Institute, Madras, 170 pp.

Ananthakrishnan, T.N. 1989. *Bioresources Ecology.* Oxford & IBH, New Delhi, 226 pp.

Ananthakrishnan, T.N. 1992. *Dimensions in Insect-Plant Interaction.* Oxford & IBH, New Delhi, 184 pp.

Ananthakrishnan, T.N. 1994. Profiles of Insect Diversity. *Curr. Sci.*, **66,** 271-280.

Ananthakrishnan, T.N. 1996. *Forest Insect Litter Communities.* Oxford & IBH, New Delhi, 174 pp.

Ananthakrishnan, T.N. 2002. *Insect, Plant and Molecular Interactions.* Madras Science Foundation, Chennai, 120 pp.

Ananthakrishnan, T.N. and Douglas Whitman, 2005. *Insect Phenotypic Plasticity.* Science Publishers, Inc., Enfield, U.S.A., 213 pp.

Arunachalam, M.K.C., Madhusoodanan Nair, Vijverberg, J. and Kortmulder, K. 1997. Food and habitat partitioning among fishes in stream pools of a South Indian river. *Int. J. Ecol. Envir. Sci.*, **23,** 271-295.

* Includes cited reference and general reference not included in text.

Avise, J.C. 1994. Molecular Markers, Natural History and Evolution. Chapman & Hall, London.

Baker, H.G. and Baker, I. 1975. Studies on nectar-contribution and pollinator-plant coevolution Pp. 100-140. *In:* L.E. Gilbert and P.H. Raven, (eds.) *Animal and Plant Coevolution* University of Texas, Austin.

Baker, H.G. and Hurd (Jr), P.H. 1968. Intrafloral Ecology, *Ann. Rev. Entomol.* **13,** 385-414.

Barraclough, J.G., Vogler, A.P. and Harrey, P.H. 1998. Resealing the factors that promote speciation. *Phil. Trans. R. Soc. Lond.*, **353,** 241-249.

Bennet, E.D. 1993. Do introduced parasitoids displace native ones? *Fla. Entomol.*, **78,** 54-63.

Coleman, D.C., Reid, C.P.P. and Cole, C.V. 1983. Biological Strategies of Nutrient Cycling in Soil system. *Adv. Ecol. Rev.*, **13,** 1-55.

Cox, G.W. and Atkins, M.D. 1979. *Agricultural Ecology*, W.H. Freeman & Co, San Francisco

Diamond, J.M. 1987. Justifying killing of birds? *Nature.* **330,** 423

Dicastri, F. and Younes, T. 1996. *Biological Diversity, Science and Development: Towards a new partnership.* CAB International, Wallingtonford, U. K.

Dinerstein, E. and Wickramanayake, E.D. 1993. Beyond hot spots: How to prioritize investments to conserve biodiversity in the Indo-Pacific regions. *Conser. Biol.*, **7,** 53-65.

Eastop, V.F. 1981. Wild hosts of aphid pests. Pp. 259-268. *In:* Thresh, J.M. (ed.) *Pests, Pathogens and Vegetation.* Pitman, London.

Ehrlich, P.R and Ehrlich, A.H. 1981. 'Extinction': the causes and consequences of disappearance of species. Random House, New York

Ehrlich, P.R. and Ehrlich, A.H. 2004. The value of biodiversity. *Ambio*, 219-226.

Ehrlich, P.R. and Raven, P.H. 1964. Butterflies and plants: a study in coevolution. *Evolution.* **18,** 586-608.

Ehrlich, P.R and Wilson, E.O. 1991. Biodiversity studies: science and policy. *Science.* **253,** 758-762

Erwin, T.L. 1982. The Tropical Forests: The heart of biotic diverstiy. *In.* Wilson, E.O and Peter, F.M (eds.) *Biodiversity*, National Academic Press, Washington, 123-129.

Erwin, T.L. 1983. Tropical Forest Canopies: The last biotic frontier. *Bull. Ent. Soc. Amer.* **29,** 14-19.

Erwin, T.L. 1991. An evolutionary basis for conservation strategies. *Science*, **253,** 750-752.

Faegri, K. and Livander Pijl 1979. *The Principles of Pollination Ecology*, Pergamon Press, New York.

Faith, D.P. 1992. Systematics and conservation: on predicting future diversity of subsets of taxa. *Cladistics*, **8,** 361-373.

Flint. M 1991. Biological Diversity and Developing Countries Issues and Opinions. ODA, 49 pp.

Forman, A. and Godron 1986. *Landscape Ecology*. John Wiley & Sons, New York.

Franklin, J.F. 1988. Structural and functional diversity in temperate forests. *In*. Wilson, E.O and Peter, F.M (eds.) *Biodiversity*, National Academy Press, Washington, DC, 166-175.

Gadgil, M. 1997. *Biodiversity*. Hornbill Publications BNHS, India.

Gadgil, M. and Meher-Homji 1986. Location of great significance of conservation in India's biodiversity. *Pre. India Acad. Sci.*, Suppl, 165-180.

Gadgil, M. and Seshagiri Rao, P.R. 1998. *Nurturing Biodiversity: An Indian Agenda*. Centre for Environmental Education, Ahmedabad, 163 pp.

Ganeshiah, K.N. and Umaashankar, R. 1998. Remote Sensing and Mapping of Biological Resources for Conservation. *Curr. Sci.*, **75,** 176.

Ganeshiah, K.N., Umaashankar, R and Bawa, K.S. 2001. Conservation of forest genetic resources of a region: combining species-centered and ecosystem-based approaches. *In*. Umaashankar, R., Ganeshiah, K.N and Bava, K.S. (eds.) *Forest Genetic Resources:* Status, Threats And Conservation Strategies. Oxford & IBH, New Delhi, 273-281.

Gaston, K.J. 1991. The magnitude of global insect species richness. *Conser. Biol.*, **5,** 283-296.

Gaston, K.J. 1992. *Biodiversity*. Blackwell, Oxford. .

Gaston, K.J. and Hudson, E. 1994. Regional habitats of biodiversity and estimates of global insect species richness. *Biodiv. Conserv.* **67**, 37-40.

Gaston, K.J. and Williams, P.H. 1993. Mapping the world's species-higher taxon approach. *Biodiv. Letters*, **1**, 2-8

Gaston, K.J., Williams, P.H., Eggketib, P. and Humphnis, C.J. 1995. Large scale patterns of biodiversity: Spatial variations in family richness. *Pro. Royal Society-London*, **260,** 149-154.

Gaston, K.J. 2000. Global patterns of biodiversity. *Nature*, **405,** 220-227.

Gaston, K.J. (ed.) 1996. *Biodiversity: A Biology of Numbers and Difference*. Blackwell Science, U. K. 396pp.

Gauld, I.D. 1999. Inventorying and monitoring biodiversity: a taxonomist's perspective. *In: Proc. Int. Sym. Med. Biodiversity*. Rome,14-16.

George, W. 1963. *Animal Geography*. Heinemann, London, 129 pp.

Glowka, L., Burhenne-Guilmin Synge, H., McNeely, J.A. and Giindling 1994. *A guide to the Conservation of Biodiversity*.

Goldfray, H.C.J. 2002. Challenges to taxonomy. *Nature*. **417,** 17-19.

Gopalakrishnan, K. 1971. The biology of the Hoogly-Matla estuarine System. (West Bengal) India with special reference to its fishes. *J. Mar. Biol. An. India*, **13,** 182-194.

Gray, J.S. 1997. Marine Biodiversity : Pattern, Threats and Conservation Needs: GESAMP studies, 6, 24, Int. Maritime. Org. London.

Groombridge, B. 1992. Global Biodiversity: Status of Earth's Living Resources. Chapman & Hall, London.

Hallmom, G.J. and Schwalbe, C.P. 2002. *Invasive Arthropods in Agriculture.* Oxford & IBH, New Delhi, 447 pp.

Harrington, R and Stork, N.E. 1995. *Insects in a changing environment.* Academic Press, London, 535 pp.

Hawkins, B.A. and Sheehan, W. 1994. *Parasitoid Community Ecology.* Oxford University Press, Oxford, 515 pp.

Hawksworth, D.L. and Ritchie J.M. 1993. Priorities for biosystematics research support in developing countries: microorganisms and invertebrates. CAB International, U.K, 73 pp.

Hirose, Y. 1994. Determinants of species richness and composition in egg parasitoid assemblages of lepidoptera. *In:* Hawkins, B.A. and Sheclan, W. (eds.) *Parasitoid Community Ecology,* Oxford University Press, 19-29.

Heywood, 1995. *Global Diversity and Assessment.* Cambridge University Press, New York.

Hora, S.L. 1947. Torrential fishes and the significance of their distribution in zoogeographical shores. *Bull. Nat. Geo. Soc. India,* Banaras, **7**, 1-10.

Hora, S.L. and Law. 1942. Respiratory adaptations of South Indian Homalopterid fishes. *J. Roy. Asiatic. Soc. Bengal,* (Science), **8,** 32-46.

Huston, M.A. 1979. A general hypothesis of species diversity. *American Naturalist,* **113,** 81-101.

Huston, M.A. 1994. Biological Diversity. The coexistence of Species on Changing Landscapes. Cambridge University Press, U. K. 681 pp.

Hutchinson, G.E, 1967. A Treatise in Limnology. Vol. II. Introduction to lake biology and the limnoplanktons, John Wiley.

IUCN 1992. Protected Areas of the World: A review of National Systems (4 Vols.) WCMC, Cambridge, and IUCN Commission on National Parks and Protected Areas. IUCN, Gland, Switzerland.

Ismail, S.1997. *Vermicology: The Biology of Earthworms,* Orient Longman, 92 pp.

Jolivet, P. 1992. *Insects and Plants: Parallel Evolution and Adaptation.* Sandhil Crane Press, Florida, 190 pp.

Kapur, A.P. 1959. Geographical variations in the colour patterns of some Indian ladybird beetles. *Proc. 1st All India Congress, Zool. Ptz.* 479-492.

Kevan, D.K. 1962. *Soil Insects.* Witherby, London.

Khorhov, T.N. 1991. *In situ* Conservation of Biological Diversity. *Biol. Edu.,* **8,** 20-26.

Kumar, H.B. 1999. *Biodiversity and Sustainable Conservation.* Oxford & IBH, New Delhi, 409 pp.

Kunte, K., Joglekar, A., Utkarsh, G. and Padmanabhan, P. 1999. Patterns of butterfly, birds and tree diversity in Western Ghats. *Curr. Sci.,* **77,** 577-586.

Lawton, J.H. and May, R.M. 1995. *Extinction rates.* Oxford University Press, Oxford.

Lawton, J. H. 2000. Excellence in Ecology: community ecology in a changing world, Ecological Institute, Oldendorf, Germany, 227 pp.

Letourneau,D.K. 1987. The enemies hypothesis: Tritrophic interactions and vegetation diversity in tropical agroecosystem. *Ecology*, **69,** 1616-1622.

Lovelock, J. 1988. *The Ages of Gaia.* W.W. Norton and Company, New york.

MacArthur, R.H. and Wilson, E.O. 1963. An equilibrium theory of insular zoogeography. *Evolution*, **17,** 373-387.

MacArthur, R.H. and Wilson, E.O. 1967. *The Theory of Island Biogeography.* Princeton University Press, New Jersy.

MacArthur, R.H. 1985. Patterns of Species Diversity. *Biol.Rev.*, **40,** 510-533.

Magurran, A.E. 2004. *Measuring Biodiversity.* Blackwell Science, London, 256 pp.

Mallet, J. 1996. The genetics of Biological Diversity: from varieties of species. *In:* Gaston. K.J. (ed.) *Biodiversity: A Biology of Numbers and Difference.* Blackwell Science, Oxford, 13-47.

Mani, M.S. 1968. Insect life in altitudinal torrential streams of Himalayas. *St John's College. Mag.* Agra, 1-7.

Mani, M.S. 1974. *Ecology and Biogeography of India.* Dr. W. Junk, N. V. Publishers, The Hague, 773 pp.

Manoharan, T.M., Biju, S.D., Nair, T.S. and Easa, P.S. 1999. *Silent Valley: Whispers of season.* Kerala Forest Department, Trivandrum, 421 pp.

Manter, H.W. 1996. Parasites of fishes as biological indicators of recent and ancient conditions. Pp. 59-71. *In:* McCauley, J.E (ed.) *Host-Parasite Relationships.* Oregon State, University Press. Coravallin

Mashelkar, R A. 2001. Intellectual property rights and the Third World. *Curr. Sci.*, **8,** 955-965.

Mattson, W.J., Lawrence. R.K., Haack, R.A., Herms, D.A. and Charles, P.J. 1988. Different strategies of woody plants against different insects feeding guilds in relation to plant ecological strategies and intimacy of association with insects. *In:* Mattson, W.J., Levienx, J. and Bernard-Ragen, C. (eds.) *Mechanism of Woody Plant Defenses against Insects.* Springer-Verlag, New York. 3-37.

May, R.M. 1994(a). Biological Diversity: Difference between land and sea. *Phil. Trans. Roy. Soc. London,* **13,** 105-111.

May, R.M. 1994(b). How many species inhabit the earth? *Sci. Amer.*, 18-24

May, R.M. and Stumpf,M.P.H. 2000. Species-Area relationship in Tropical Forests. *Science,* **290,** 2084-2086.

Mayr, R.E. 2001. *What Evolution Is.* Basic Book. 318 pp.

McNeeley, G.A., Miller, K.R., Reid, W.V., Mittermeir, R.A. and Werner, T.R. 1990. *Conserving the World's Biological Diversity.* (IUCN).

McNeely, J.A. 2002. The role of taxonomy in conserving biodiversity. *J. Nat. Conser.*, **10,** 145-153.

Measey, G.J., Gowev, D.J., Oommen, O.V. and Wiltinom, M. 2003. Quantitative surveying of endogeic vertebrates – A case study of *Gegenophis ramaswamii*

(Amphibia : Gymnophora : Caeciliidae) in South India. *Appl. Soil. Ecol.* **23**:43-53.

Mittermier, R.H., Myers, N., Gill, P.R. and Mittermier, C.G. 1999. *Hotspots: Earth's richest and most endangered terrestrial ecoregions.* Agrupacion Sierra Madre, Mexico.

Miller, R.R. 1958. Origin and affinities of the fresh water fish fauna of western north america. *In:* Hubbs, C.L. (ed.) *Zoogeography.* Publ. 51, 187-222, Washington DC., *Amer. Asso. Airan. Science.*

Mohan, M.S.S., Sudha Nair and Ranjit Daniel, R.J. 1995. Bioindicators: Identification and use in assessing environmental disturbances. MSSRF, 26 pp.

Moritz, T. 2002. Building the biodiversity communities. *D-Lib.magazine.* **8,** 1-7.

Mushtak Ali, T.M. and Ganeshiah, K.N. 1998. Mapping diversity of ants and root grubs. *Curr. Sci.*, **75,** 176.

Myers, M., Mittermier, R.A., Mittermier, C.G., Gustaro A.B da Fonesca and Kent, J. 2000. Biodiversity hot spots for conservation priorities. *Nature,* **403,** 853-858.

Myers, N. 1988. Threatened "hot spots" in tropical forests. *The Environmentalist,* **8,** 1-20.

Myers, N. 1996. *Environmental Services of Bioodiversity.* PNAS (USA) **93**, 2764-2769.

Myers, N. 1997. Consumption in relation to population, environment and development. *The Environmentalist,* **17,** 34-44.

Myers, N. 1998. Threatened biotas: "Hotspots" in Tropical Forest. *The Environmentalist,* 187-208.

Myers, N. 2000. The new millennium: A ecology and an economy of hope. *Curr.Sci.,* **78,** 686-693.

Nadkarni, N.M. 1994. Diversity of species and interactions in the upper canopy of forest ecosystem. *Amer. Zool.,* **34,** 70-78.

Naess, A. 1986. *Ecology, Community and Lifestyle.* Cambridge University Press, Cambridge.

Narayanan, L.R.A. 1999. *Remote Sensing and its Application.* University Press, Hyderabad, 215 pp.

National Research Council 1987. *Agricultural Biotechnology.* National Academy Press, Washington D C, 209 pp.

Norse, E.A. and McManus, R.E. 1980. Ecology and living resources biological diversity. *In.* Environmental Quality: *The Eleventh Annual report of the council on Environmetal Quality.* Washington, 31-80

Noss, R.F. 1990. Indicators for monitoring biodiversity: A hierarchical approach. *Conser. Biol.,* **4,** 355-364.

Noss, R.F. and Cooperrider, A.Y. 1994. *Saving Nature's Legacy. Protecting and Restoring Biodiverstiy.* Island Press, Washington D C. 416 pp.

Perrings, C. 1995. Economic values of biodiversity, In: *Global Biodiversity Assessment.* Cambridge University Press, Cambridge, 823-914.

Primack, R.B. 2002. *Essentials of Conservation Biology*. Sinauer Associates, U.S.A. 698 pp.

Pushpangadan, P. 2002. Biodiversity and Emerging Benefit Sharing Arrangements: Challenges and Opportunities for India. *PINSA*, **B68,** 297-314.

Pushpangadan, P. and Nair, K.V. 2001. Future of systematics and biodiversity research in India: Need for a National comortium and National agenda for systematic biology research. *Curr. Sci.*, **80**, 631-638.

Pushpangadan, P., Ravi, K. and Santhosh, V. 1997. *Conservation and Economic Evaluation of Biodiversity*. Oxford and IBH., New Delhi.

Rabinowitz, D., Cairns, S. and Dillon, J. 1986. Seven forms of rarity and their frequency in the flora of the British India. *In:* Sonte, M.E. (ed.) *Conservation Biology: The Scarcity and Diversity*. Sinauer Association, Sunderland, Min, U.S.A.

Ramakrishnan, P.S. 2002. Coping with uncertainties: Role of Traditional Ecological Knowledge in Socio-Ecological Landscape management. *PINSA*, **B68**, 235-254.

Ramakrishnan, P.S. 2004. *Ecology and Sustainable Development*. Nahmal Book Trust, New Delhi, 198 pp. (First reprint).

Rana, R.S. 1991. Biodiversity and Crop improvement *ex situ* conservation. *Biol. Edu.*, **8,** 27-33.

Ranganath, H.A. 2002. Evolutionary biology of *Drosophila nasuta* and *Drosophila albomicans*. *PINSA*. **B68,** 255-272.

Reid, W.V. 1992. Conserving life's diversity. *Environ. Sci. Technol.*, **26,** 1090-1095.

Reid, W.V. 1998. Biodiversity hot spots. *Tree*, **13,** 257-279.

Richards, O.W. 1961. Insect polymorphism, *Symposum I, Royal Entomological Society*, London.

Robert, C.M., McClean, C.J., Veron, J.E.N., Hawkins, J.P., Allen, G.R., McAllister, D.E., Mittremeir, C.G., Scheuler, F.W., Spalding, M., Wells, F., Vjune, C. and Werner, T.B. 2002. Marine biodiversity hotspots and conservation priorities for tropical reefs. *Science*. **295**,1280-1284.

Rosenzweig, M.L. 1995. *Species Diversity in Space and Time*, Cambridge University Press, Cambridge.

Samways, M.J. 1994. *Insect Conservation Biology*. Chapman & Hall, London, 358 pp.

Savage, J.M. 1995. Systematics and the biodiversity crisis. *Bioscience*, **45,** 673-678.

Sathish Chandran Nair 1991. The Southern Western Ghats: A Biodiversity Conservation Plan. INTACH, New Delhi, 92 pp.

Schluter, D and Ricklefs, R.E. 1993. Species Diversity: *An Introduction; In*: Ricklefs, R.E. and Schluter, D. (eds.) *Species Diversity in Ecological Communities*. Chicago University Press, Chicago.

Sessiour, G. 1995. *Deep Ecology for the 21st Century*. Shambala Books., Boston.

Seasted, T.R. 1984. The role of microarthropod in decomposition and mineralisation process. *Ann. Rev. Entomol.*, **29,** 25-49.

Selvam, V. 2003. Environmental classification of mangrove wetlands India. *Curr. Sci.*, **84,** 757-764.

Singh, J.S. 2002. The biodiversity crisis: a multifaceted review. *Curr. Sci.*, **82,** 638-647.

Singh, J.S. and Khurana, E. 2002. Paradigms of biodiversity: An Overview, PINSA, **B 68**, 273-296.

Solbrig, O.T. 1991. From Gene to Ecosystem: A Research Agenda For Biodiversity. IUBS, Paris.

Soule, M.E. ans Wilcox, B.A.1980. Conservation Biology: An Ecological-Evolutionary Perspective. Sinauer.

Southwood, T.R.E. 1976. Bionomic strategies and population parameters. Pp. 26-48. *In:* May, R.M.(ed.) *Theoretical Ecology: Principles and Practices.* Saunders, Philadelphia.

Stork, N.E. 1988. Insect diversity: facts, fictions and speculations. *Biol. J. Linn. Soc.*, **35,** 321-337.

Strong, D.R. 1997. Fear no weevil? *Science.* **277,** 1058-1059.

Subbaram, N.R. and Thyagarajan, G. 1998. Gearing up for the efficient management of Intellectual Property Rights in the 21st Century. COSTED Occ. Paper Series, 92 pp.

Subramanian, K.A. and Sivaramakrishnan, K.G. 2005. Habitat and microhabitat distribution of stream insect communities of the Western Ghats. *Curr. Sci.* **89(6):** 976-987.

Swaminathan, M.S. 1991. *Biodiversity in Agriculture.* Macmillan India, Madras, 371 pp.

Swaminathan, M.S. and Jana, S. 1992. *Biodiversity.* Macmillan, India, Madras, 326 pp.

Swaminathan, M.S. 2000. Government-Civil society partnerships in integrated gene management. *Curr. Sci.*, **78,** 555-562.

Swaminathan, M.S. 2002. The past, present and future: Contribution of farmers to the conservation and development of genetic diversity. Pp. 23-31. *In:* Engels, J.M.M. (ed.) *Managing plant genetic diversity* IPGRI.

Terborgh, J. 1992. *Diversity and The Tropical Rain Forest.* Scientific American, New York.

Thompson , J.N. 1994. *The Coevolutionary Process.* The University of Chicago Press. Chicago.

Tyagi, B.K. 2003. *Medical Entomology.* Scientific Publishers (India), Jodhpur, 262 pp.

Tyagi, B.K. 2004.*The Invincible Deadly Mosquitoes.* Scientific Publishers (India), Jodhpur, 267 pp.

Tzschupke, W. 1998. Forest sustainability-A contribution to conserving the basis of our existence? *Pl.Res.Develop.* **47/48**, 14-28.

Umashankar, R., Ganeshiah, K.N. and Bawa, S.K. 2001. *Forest Genetic Resources: Status, Threats and Conservation Strategies.* Oxford & IBH, New Delhi.

UNEP (United Nations Environmental Programme) 1992. *Conservation on Biological Diversity.* UNEP, Nairobi.

Urban, D.L., O'Neill, R.V. and Shugart, H.H. 1987. Landscape ecology. *Biosciences.* **37,** 119-127.

Utkarsh, G., Joshi, N.V. and Gadgil, M. 1998. On the patterns of Tree diversity in the Western Ghats of India. *Curr. Sci.*, **75,** 594-601.

Van Emden, H.F. 1981. Wild plants in the ecology of insect pests. Pp. 251-261.*In: Pests, Pathogens and Vegetation* (ed.) J.M. Thresh, Pitman, London.

Van-Wright, R.I. 1992. Systematics and Biodiversity. *In.* Groombridge, B (ed.) *Global Biodiversity-Status: The Earth's Living Resources.* Chapman & Hall, London: 7-12.

Virchow, D. 1998. *Conservation of Genetic Resources.* Springer-Verlag, Berlin.

Walker, B. 1992. Biodiversity and ecological redundancy. *Couserv. Biol.* **6**, 18-23.

Wallwork, J.A. 1970. *Ecology of Soil Animals.* McGraw Hill, New Delhi.

Way, M.J. and Cammell, M. 1981. Effects of weeds and control on invertebrate pest ecology. *In:* Thresh J.M. (ed.) *'Pest, Pathogen and Vegetation'*, Pitman, London, 443-458.

WCMC (World Conservation and Monitoring Centre) 1992. *Global Biodiversity: Status of the Earth's Living Resources.* Chapman & Hall, London.

Willcox, B.A. and Murphy, D.D. 1985. Conservation strategy: the effects of fragmentation on extinction. *Amer. Nat.,* **125,** 879-887.

Wilson, E.O. and Bossert 1971. *A Primer of Population Biology.* Sinauer Assoc.Inc. Connecticut, USA, 192 pp.

Wilson, E.O. 1984. *Biophilia: The Human Bond with Other Species.* Harvard University Press, Cambridge M A.

Wilson, E.O. and and Peters, F.M. 1988. *Biodiversity.* National Academy Press., Washington, D C.

Wilson, E.O. 1992. *The Diversity of Life.* Belknap Press, Harvard University.

Wilson, E.O. 1997. Introduction Pp. 1-6 In: M.L. Reaka-Kundla *et al.* (eds.) *Biodiversity II Understanding and Protecting our Biological Resources.* Joseph Henry Press., Washigton, D C. USA., 551 pp.

WRI (World Resource Institute) 1992. *Towards Sustainable Development.* Oxford University Press, New York.

WRI, IUCN and UNEP. 1992. *Global Biodiversity Strategy,* 244 pp.

Glossary

Agroecology	Holistic study of agroecosystem
Agroecosystem	An ecosystem comprise of agricultural activities
Allozyme	A subject of isozymes which represent variants of polypeptides
Alpha diversity	Species diversity within a habitat.
Amplification	Products of large amounts of a small segment of a DNA molecule.
Autecology	Study of ecology of a single species.
Beta diversity	Species diversity between habitats or along an environmental gradient.
Bioaccumulaiton	Organism that accumulate particular substances within their tissues.
Biodiversity	The variety of life and its processes including their genetic differences, the communities and ecosystem in which they occur & functional dynamics including evolutionary processes.
Biogeography	The study of geographic distribution of organisms.
Biological control	The use of natural enemies to control pests.
Biomarkers	Biochemical changes in an organism caused by chemicals.
Biomass	The totality of organism, plant or animal.
Biome	A large, regional ecological unit generally defined by a collection of organisms, plant or animal.
Biomonitors	Species in which changes in numbers or distribution are studied over time.
Biophilia	Innate love for nature and natural diversity. Coined by E.O. Wilson.
Bioprobes	Symptom-oriented biological effect test system.
Biosphere reserve	A core reserve surrounded by buffer zones.

Biota	All organisms found in a given area.
Biotechnology	Techniques that use living organism or part of organism, to make or modify products to improve plants, animals or to develop microorganisms for specific uses.
Clone	A population of individuals all derived asexually from the same single parent.
Coevolution	The parallel evolution of two kinds of organisms that are interdependent.
Colonization	The immigration of species into a new habitat and founding of a population.
Conservation biology	Studies concerned with the dynamics of diversity, scarcity and extinction.
Corridor	A route that allows movement of individuals or taxa from one region or place to another.
Cryopreservation	Storage of living materials at very low temperature, usually in liquid nitrogen.
Desertification	The impoverishment of arid, semi arid and sub humid ecosystem by human influence.
Diversity	Ecological measures of the number of species and their relative abundance in a community.
Drilosphere	Entire volume soil and litter and the associated earthworms.
Ecosystem management	Management of ecosystem that seeks to protect viable population of all native species to prevent long term ecological degradation.
Ecosystem	A dynamic complex of plant, animal and microorganism communities and their associated nonliving environment.
Endemism	Number of species restricted to a particular geographic area or region.
Environmental gradients	Change in ecological or environmental features across space such as elevation, moisture , temperature *etc*.
Exotic species	Species accidentally or deliberately introduced by man.
Ex-situ conservation	Conservation outside its original or natural habitat.
Fitness	The relative survival value and reproductive capability of a given genotype in comparison with others in the population.
Flagship species	Species which attract popular support for their conservation because they are popular and charismatic.
Fragmentation	The breaking up of a large habitat/ ecosystem/landscape into more than one bit.

Gaia	The view that the earth is a super organism refers to the feedback mechanisms whereby biological processes modify their physico-chemical conditions appropriate to biological process.
Gamma (γ) richness	Number of species found within a large region region which includes several habitats.
Gap analysis	Identification of spatial gaps in species protection and management programmes by remote-sensing data.
Gap formation	The generation of patches in a landscape.
GATT	General Agreement on Tariffs and Trade.
Gene bank	Facility where germplasm is stored as seeds, pollen *in vitro* culture.
Gene ecology	Study of population genetics in relation to habitats.
Gene pool	Total genetic material of a freely interbreeding population.
Generalists	Species that have a broad habitat range.
Genetic drift	Gene frequency changes occurring randomly in small populations due to chance.
Genetic erosion	Reduction of the genetic diversity of a species due to environmental changes or human intervention.
Genome	The genetic make-up of an organism *ie.,* sum of all genes of an organism.
GIS	A computerized system of organization and analysing any spatial data and information (Geographical Information System).
Habitat fragmentation	Process by which habitats are increasingly subdivided into small units.
Hot spot	A specific place inhabited by an organism characterized by unusual high species richness.
In-situ conservation	Conservation of plant and animals in the wild or in farmer's fields.
Indicator species	A species used as a guage for the condition of a particular habitat.
Intellectual property rights (IPR)	Protection of an invention through the use of legal methods such as patenting, copy rights and plant breeder's rights.
Keystone species	A species that plays a vital role in an ecosystem and upon which a large part of a community depends.
LANDSAT	One of the several satellites used to generate maps through remote sensing.
Land races	Farmer developed cultivars or crop plants which are adapted to local environmental conditions.

Landscape	A heterogenous land area comprising a cluster of interacting ecosystem.
Life form	Groups of species that use similar habitat for feeding and breeding and other needs.
Litter	Upper layer of organic material on the soil surface.
Megadiversity countries	Countries with very great species diversity.
Molecular marker	A gene or gene product of known chromosomal location that marks another gene or selected chromata.
Mutualism	An interspecific interrelationship benefiting both organisms.
Mycorrhiza	Symbiotic association of a fungus with the roots of higher plants.
Networking	Establishing and strengthening links between individuals or groups with similar objectives.
Niche	Place in a habitat specifically used by a species.
Nutrient cycling	Flow of nutrients through a large ecosystem and enabling reusing them.
Off-site conservation	*Ex-situ* conservation
Oligotropic	Lakes with low productivity.
On-site conservation	*In-situ* conservation.
Pangaea	The Super continent of the Permian era.
PCR (Polymorphic chain reaction)	A method for making multiple copies of fragments of DNA.
Phenetic	Pertaining to phenotypic similarities.
Phenotype	The observable characters of an organism, resulting from the genotype.
Plasmid	A genetic element that has an independent existence apart from the main DNA in the cell.
Plasticity	A genetically determined but environmentally induced variations in any character of an organism.
Pleiotropism	How a gene may affect several aspects of the phenotype.
Ramsar conservation	An international convention established for the purpose of identification, conservation and management of threatened wetlands of the world.
Rapid assessment in programme	Quick collection, analysis and dissemination of information areas potentially important from the point of view of biodiversity.
Regional reserve network	A set of native reserves functionally connected that provide greater protection than individual reserves.

Remote sensing	Techniques for analysing landscape patterns using satellite imagery.
Restoration ecology	Use of ecological principals to turn a degraded ecosystem to its original state.
Restriction enzyme	Enzyme produced by bacteria used to assess genetic diversity of organisms.
RFLP (Restriction fragment length polymorphism)	The variation within a population, of the lengths of restricted fragments formed by breaking DNA with a restriction enzyme.
Seed bank	A facility where the seeds are stored for ex-situ conservation.
Species abundance	Number of individuals of a species in a defined region.
Species diversity	The number of different species present in a defined area.
Species evenness	Equitability of different species or genera by their relative abundance in a particular region.
Species turnover	Changes in species richness between areas.
Species-area relationship	The number of species per area varies with the size of the area-greater the area, larger the species numbers.
Sustainable development	Development to meet the needs and aspiration of both current and future generations.
Sympatric	Two or more populations of species that occur in same geographic area.
Taxonomy	The theory of practice of classifying organism.
Threatened species	Species at risk of extinction.
Transgenic	An organism, which has received through biotechnological methods specific genes or group of genes from another organism.
TRIPS	Trade related aspects of Intellectual Property Rights.
Umbrella species	Species that require large areas to maintain viable populations and if such species are conserved many other species will be incidentally protected.
Use value	Values obtained through the actual use of a biodiversity element.
Viable population	Population that contains adequate number of individuals, well distributed, to ensure long term survival.
Vulnerable species	Species sensitive to human activity or a species which though not endangered, but a poses a risk of extinction.
WIPO	World Intellectual Property Organisation.
WIPONET	A global Intellectual Property network set up by WIPO.
World heritage site	A type of protected area historically important and recognised by UNESCO.

Index